Study Guide

Gary Wisehart
Michael J. Leboffe
San Diego City College

to accompany

Biology: Exploring Life

SECOND EDITION

Gil Brum
California State Polytechnic University, Pomona

Larry McKane
California State Polytechnic University, Pomona

Gerry Karp
Formerly of the University of Florida, Gainesville

JOHN WILEY & SONS, INC.

New York / Chichester / Brisbane / Toronto / Singapore

Preface

Study Guide Organization

Although many students anticipate that their first biology class will cover material with which they are already partially familiar, they often find biology to be quite challenging and foreign. There are many new terms and concepts to master, and it's often difficult to know where to begin studying. This Study Guide contains questions that direct your review of important definitions and concepts covered in your text. It also has exercises that give you critical thinking practice by finding relationships between concepts.

Study Guide chapters correspond to chapters in *Biology: Exploring Life*, Second Edition by Brum, McKane and Karp. Each includes the following components.

- Section Concept Maps - These relate each chapter to other relevant chapters in the text and to text themes (F/F - Form and Function, AUE - Acquiring and Using Energy, E/A - Evolution and Adaptation, BO, R, H - Biological Organization, Regulation, and Homeostasis, and U/D - Unity within Diversity). Each should be studied prior to reading the chapter to put it into perspective with what has already been covered.

- Go Figure! - These are questions designed to clarify and expand upon principles presented in selected text figures. The intent is to focus attention on the value of these pedagogical aids. In some cases the answer may be determined by examining the figure closely. In others, the answer requires interpretation of the figure or application of what you have learned elsewhere.

- Matching Questions - These test your recall of definitions. Most items are from the Key Terms list at the end of each chapter. Use these questions as a review after reading the chapter and attending lecture. Answers to odd-numbered items are in the Appendix; answers to all questions are in the *Instructor's Manual* available to your professor.

- Multiple Choice Questions - These test your recall of facts and concepts covered in the chapter and also should be used as a review after reading and lecture. Questions are arranged in pairs testing comparable material. The answer is given to the first (odd-numbered) question of the pair. Occasionally, sequences of two or more questions are paired. These sequences are clearly marked and answers are provided to all questions of the first sequence. Answers to all questions are in the *Instructor's Manual* available to your professor.

- Concept Map Construction Problems - These allow you to develop your concept mapping skills. They help you recognize relationships between concepts in the current chapter and between concepts in the current chapter and previous chapters. Because they are comprehensive, attempt these questions as a final review of the material before moving on to the next chapter. Your professor has been provided with a sample concept map for each question in the *Instructor's Manual*. Concept map construction is explained on the next page.

- Answers to Odd Numbered Problems - These include multiple choice questions only. Answers to all matching and multiple choice questions, and sample solutions to concept map problems have been provided for your instructor in the *Instructor's Manual*.

We hope this Study Guide helps you in your biology class. Good luck!

Concept Map Construction

Suppose you were asked to describe how human, gasoline, motor boat, truck, and vehicle are related. There are lots of ways you could set about doing this. For example, you could write a narrative description, give an oral presentation, or draw each object to illustrate how they interact. Another familiar method is an outline. One possible outline is shown below.

    I. Vehicles
        A. Water Vehicles
            1. Sail Boat
            2. Motor Boat
        B. Land Vehicles
            1. Examples
                a. Truck
                b. Bicycle
            2. Wheels
  II. Power Source
        A. Human
        B. Engine
            1. Electric
            2. Gasoline
        C. Wind

An outline has advantages over the other methods. In a short time, someone can examine it and understand the hierarchical relationships visually reinforced by indentations. But notice many important connections are not indicated. For instance, trucks and bicycles both have wheels, and trucks and motor boats both have engines. Neither is indicated by the outline. Further, it is unclear what power source is used by each vehicle. Because of these shortcomings, other methods have been devised to show relationships which also require a short period of time to understand and visually reinforce associations. One of these is the concept map.

Concept maps are a useful tool for showing relationships between items that might be overlooked or too cumbersome to show in an outline. The idea of a concept map is simple. There are four parts:
    1) boxed terms and/or concepts,
    2) connecting lines,
    3) "connector" words or phrases,
    4) cross links.

A hierarchy is formed by placing the broadest (most inclusive) term/concept at the top of the map and more specific terms/concepts towards the bottom. Lines are then used to indicate a relationship between boxed terms/concepts. A brief connector phrase is written over the line to let the reader know what that relationship is. If the relationship is between items in different branches, an arrow is used to indicate the direction of the relationship. Because of their flexibility in showing connections and their hierarchical arrangement, they reflect how information probably is stored in our memory.

Linking terms/concepts is often the most difficult part for beginning concept mappers. Connectors should be non-technical and usually indicate physical (under, near, connected to), temporal (precedes, follows), or logical (causes, produces) relationships, or subcategories (includes, demonstrates, such as). Keep them simple and do not include words or phrases which could be incorporated as boxed terms/concepts.

On the next page is a sample of a concept map developed from the following terms: human, gasoline, motor boat, truck, and vehicle.

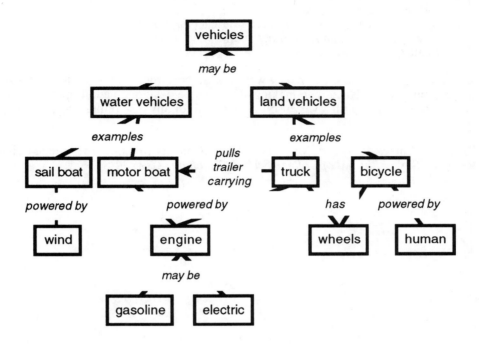

Points of Interest
1.   Notice that the term "vehicle" was assigned, but the plural "vehicles" is used in the concept map. Terms will be assigned in the singular form, but it is acceptable to substitute with a plural form.
2.   Start at the top of the map and read down any branch. Notice that each pair of terms joined by a line and a connector phrase makes sense. Each pair must stand alone as a sensible statement.
3.   Additional terms (water vehicles, land vehicles, bicycle, wind, sail boat, engine, and wheels) were used to clarify and expand relationships. You are encouraged to do this, and it may even be essential to include these intermediate terms to "bridge the gap" between concepts that are not closely related. *Rest assured, however, that any term can be incorporated into any concept map with the appropriate connectors and intermediate concepts.*
4.   Notice that lines extending from an item using the same connector phrase arise from the same point on the rectangle (*e.g.*, "water vehicle" connected with "sail boat" and "motor boat"). Lines extending from an item using different connector phrases arise from different points on the rectangle (*e.g.*, "bicycle" connected with "wheels" and "human"). This makes reading the map easier.
5.   Notice the cross link indicated by the arrow between "truck" and "motor boat." The relationship only works in one direction (*i.e.*, trucks pull motor boats, not the other way around), so the arrow is essential.

Don't be concerned about constructing the "correct" map; many are possible for any given set of terms. Since there is no single correct answer, other criteria must be used for determining the quality of your map. Often the logic of the hierarchical arrangement, number and quality of added terms, use of cross links, and validity of connector phrases are criteria for evaluation. Mechanical aspects may also be used. These include use of rectangles around concepts, connector phrases, and arrows for cross links. Your professor will let you know how and if yours will be evaluated.

Wallace, Mintzes and Markham (1992)* suggest several "do's" and "don'ts" of concept mapping that our students have found useful. These are:

1. Don't become discouraged; do set it aside prior to the frustration point.
2. Do work with other students.
3. Do seek assistance from your professor or teaching assistant.
4. Do construct concept maps for other courses.
5. Do practice, practice, practice.

Completed concept maps give the illusion they were easy to construct. Our experience has been that novice mappers struggle, but with practice, map making skills can be developed and improved. Once developed, concept mapping provides you with a very useful study tool for any discipline, not just biology.

*Wallace, J.D., J.J. Mintzes, and K.M. Markham. 1992. Concept Mapping in College Science Teaching - What the Research Says. *Journal of College Science Teaching* 22(2):84-86.

Table of Contents

Chapter 1

Biology: Exploring Life

Section Concept Map

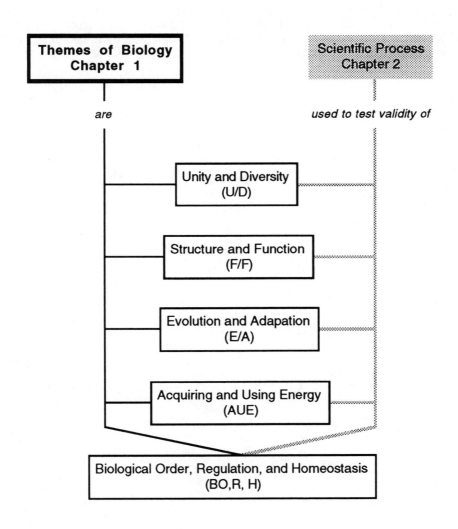

Go Figure!

Figure 1-4:   New discoveries in science often result from the introduction of new technologies. For example, in the 1880s molten materials, including wax, began to be used to infiltrate biological specimens so that thin slices could be made for examination under a microscope. This technology revolutionized the field. Which levels of biological organization do you think were unknown to science until new technological innovation revealed their presence?

Figure 1-5:   This figure implies many historical relationships. What does it imply about the kingdom's historical sequence? Which type of organism has been on earth the longest, fern or conifer? What does the caption imply about the relationship between number of species in a kingdom and the diversity of species within a kingdom?

Figure 1-10:   Arrange these birds into four groups so that each bird in a group is more closely related by ancestry to all other birds in the group than to a bird in another group. Groups of a single bird are acceptable. Do the groups reflect similarity in habitat as well as degree of evolutionary relationship?

Matching Questions

Write the letter of the phrase that best matches the numbered term on the left. Use each only once.

_____ 1.   biology

_____ 2.   cell

_____ 3.   unicellular

_____ 4.   metabolism

_____ 5.   energy

_____ 6.   gene

_____ 7.   development

_____ 8.   enzyme

_____ 9.   homeostasis

_____10.   stimuli

_____11.   atom

_____12.   biosphere

_____13.   population

_____14.   proton

_____15.   binomial

_____16.   system of classification

_____17.   family

_____18.   Monera

_____19.   DNA

_____20.   adaptation

a.   all chemical reactions within an organism

b.   a unit of hereditary instruction

c.   the smallest organization level of life

d.   a change in form during one individual's life time

e.   the science of life

f.   changes in the environment

g.   the non-material property used by organisms to do the work of building and maintaining structure

h.   smallest unit of a substance having the properties of that substance

i.   used to describe an organism that consists of one cell

j.   all members of a species living in the same habitat

k.   groups objects according to similarities

l.   includes every place on earth where living organisms are found

m.   the taxonomic category composed of similar genera

n.   kingdom name

o.   the hereditary molecule

p.   substance which increases the rate of a specific chemical reaction

q.   a subatomic particle

r.   maintenance of internal conditions within a limited range

s.   increases the likelihood that an organism will reproduce

t.   consisting of two names

Multiple Choice Questions

1. When the authors of your text say that living organisms are complex, they mean
   a. living things are composed of many different chemical types.
   b. living things are composed of very many chemical types arranged in sophisticated and precise ways.
   c. living things are composed of very many chemical types arranged in simple repeating patterns.
   d. living things are composed of very many chemical types arranged in a uniform pattern.
   e. living things are able to process and use energy.

2. Which is the least complex level of organization capable of performing simple biological functions?
   a. cell.
   b. population.
   c. molecule.
   d. atom.
   e. community.

3. Which IS NOT used to characterize living things?
   a. high degree of complexity
   b. composed of one or more cells
   c. maintains a constant temperature
   d. grows in size and changes in appearance
   e. responds to changes in the environment

4. Which IS NOT used to characterize living things?
   a. produce offspring similar to themselves
   b. inherit genes from parents
   c. grows in size and changes in appearance
   d. uses sunlight directly as a source of energy
   e. maintains homeostasis

5. Nearly two million species on earth have been identified. The greatest number of these are
   a. unicellular.
   b. multicellular.
   c. unable to carry on metabolism.
   d. unable to maintain homeostasis.
   e. prokaryotes

6. Nearly two million species on earth have been identified. Historically, the first to appear were
   a. unicellular.
   b. multicellular.
   c. unable to carry on metabolism.
   d. unable to maintain homeostasis.
   e. eukaryotes

7. The original source of energy used by the vast majority of living things is
   a. chemicals found in plants.
   b. chemical compounds found in soil.
   c. chemicals found in bacteria.
   d. chemicals found in food.
   e. sunlight

8. Members of the Animal Kingdom directly depend upon _____ as an energy source.
   a. soil minerals
   b. sunlight
   c. water
   d. tissues of other living things
   e. oxygen

9. The genetic instructions inherited from our parents consists of a large collection of
    a. preprogrammed behavioral responses.
    b. preformed body parts.
    c. miniature internal organs.
    d. genes.
    e. cell organelles.

10. The genetic instructions inherited from our parents are chemically composed of
    a. enzymes.
    b. energy.
    c. preformed body parts.
    d. DNA.
    e. cell organelles.

11. As a tadpole transforms into an adult frog, there is very little if any change in size, but there is a great deal in its external appearance. This modification is an example of
    a. reproduction.
    b. growth.
    c. development.
    d. metabolism.
    e. homeostasis.

12. In most species, as a tadpole transforms into an adult frog, its diet switches from plants to animals. This switch is accompanied by a change in the type of enzymes present in the digestive tract. This modification is an example of
    a. reproduction.
    b. growth.
    c. development.
    d. metabolism.
    e. homeostasis.

13. Organelles are composed of
    a. cells.
    b. other organelles and cytoplasm.
    c. molecules.
    d. populations.
    e. communities.

14. Cells are composed of
    a. cells.
    b. organelles and cytoplasm.
    c. molecules.
    d. populations.
    e. communities.

15. Which is the smallest level of organization capable of performing "higher" biological functions including sight?
    a. cell
    b. population
    c. multicellular organism
    d. atoms
    e. community

16. Which is the smallest level of organization capable of evolution?
   a. cell
   b. population
   c. multicellular organism
   d. atom
   e. community

17. A newly discovered species is given a "scientific name" composed of its
   a. genus and species names.
   b. family and species names.
   c. order and class names.
   d. family and genus names.
   e. family and order names.

18. The system of classification in use today assigns species into taxonomic categories based upon
   a. an alphabetical arrangement of scientific names.
   b. alphabetical arrangement of species names.
   c. evolutionary relationships between species.
   d. similarity in habitat.
   e. ability to maintain homeostasis.

19. Which is the correct listing of taxonomic categories from large to smaller without skipping one of the categories given in your text?
   a. Kingdom, Phylum, Class, Order
   b. Kingdom, Genus, Family, Species
   c. Phylum, Order, Genus, Kingdom
   d. Phylum, Order, Family, Species
   e. Kingdom, Phylum, Class, Species

20. Which is the correct listing of taxonomic categories from large to smaller without skipping one of the categories given in your text?
   a. Kingdom, Genus, Species
   b. Domain, Kingdom, Species
   c. Phylum, Class, Species
   d. Family, Genus, Species
   e. Family, Order, Species

21. According to Figure 1.6, which Kingdom is directly ancestral to members of Kingdom Fungi and the Animal Kingdom?
   a. Monera
   b. Protista
   c. Plants

22. According to Figure 1.6, which organisms are most closely related evolutionally?
   a. slime molds and brown algae
   b. brown algae and red algae
   c. spiders and insects
   d. ribbon worms and earthworms
   e. fish and sea stars (starfish)

23. Two organisms live in the same place, but are members of different populations; this implies that they are
   a. different species.
   b. members of different ecosystems.
   c. the same species.
   d. members of different communities.
   e. in the Kingdom Protista.

24. Two organisms live in different places, but have the same scientific name; this implies that they are
    a. different species.
    b. members of different families.
    c. members of different genera.
    d. members of different communities.
    e. in the Kingdom Protista.

25. You possess a four chamber heart, while the heart of a frog has three. As a consequence, your heart more efficiently delivers oxygen rich blood to muscles allowing you to maintain vigorous exercise for a longer period of time than the frog. This increases your chance of survival and is an example of
    a. an anatomic adaptation.
    b. a physiologic adaptation.
    c. a behavioral adaptation.
    d. why frogs are poorly adapted.
    e. why natural selection has not favored frogs.

26. In most species, as a tadpole transforms into an adult frog, its diet switches from plants to animals. This switch is accompanied by a change in the type of enzymes present in the digestive tract. This modification increases the chance of survival by assuring the adults and tadpoles do not compete for food resources and is an example of
    a. an anatomic adaptation.
    b. a physiologic adaptation.
    c. a behavioral adaptation.
    d. why frogs are poorly adapted.
    e. why natural selection has not favored frogs.

27. Natural selection acts
    a. in a random manner upon all individuals in a population.
    b. upon variability with a population.
    c. equally upon all individuals in a population.
    d. by increasing the reproductive rate for an entire population.
    e. by decreasing the reproductive rate for an entire population.

28. Darwin proposed that, since all species produce more individuals than can survive,
    a. individuals in the population are variable.
    b. some individuals are better adapted than others.
    c. a population changes over time.
    d. some individuals in the population will not survive.
    e. offspring inherit the traits of their parents.

Concept Map Construction

Construct a concept map for each group of terms. Be sure to include appropriate connector phrases. You may add other terms as necessary and use terms in the plural or singular form..

1. biology, cell, unicellular, energy, reproduction,
2. stimulus, homeostasis, atom, cell, population
3. population, reproduction, variability, natural selection, Thomas Malthus

Chapter 2

The Process of Science

Section Concept Map

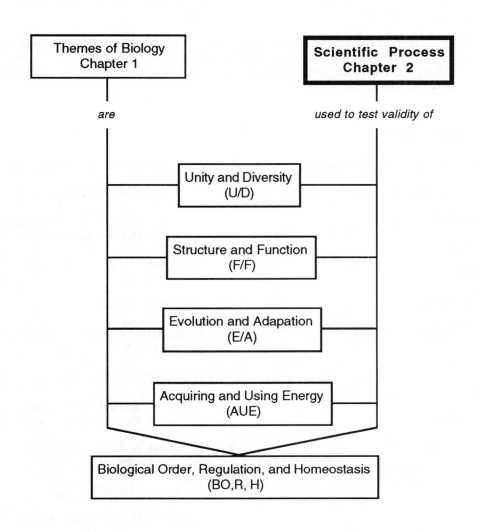

Go Figure!

Figure 2-1:   The two jars in part "a" differ by more than one variable. Besides presence of adult flies and free exchange of air between the inside and outside of the jar, what other possible differences exist? What would you conclude if after several days an open jar did <u>not</u> have maggots? Would it be acceptable procedure to keep the wax sealed jar out of the sun so that the wax would not melt, while keeping the uncovered jar in the sun?

Figure 2-2   Why was the neck of the flask bent? What was the purpose of heating the broth?. What would you conclude if no microbes grew after tipping the flask?

Figure 2-4   Why was such a large test population used? What do you think would be the best way to assign rats from the test population to either the experimental or control group? What is the experimental variable? How would you decide whether the incidence of cancer was high or normal?

Figure 2-6   Assume you are applying this scheme to the red dye #2 problem in your text. List all resources you can think of that could be used to complete the first step. Using this scheme, what action would you take if the results of an experiment were different than you hypothesis predicted?

Matching Questions

Write the letter of the phrase that best matches the numbered term on the left. Use each only once.

_____ 1.   spontaneous generation

_____ 2.   hypothesis

_____ 3.   hypothesis test

_____ 4.   variable

_____ 5.   control group

_____ 6.   experimental group

_____ 7.   data

_____ 8.   theory

_____ 9.   *Penicillium*

_____10.   carcinogen

a.   a condition which can change

b.   in an experiment, those objects exposed to the variable being tested

c.   a tentative, testable statement of cause and effect

d.   a genus of mold

e.   nonliving material producing living organisms

f.   cancer causing agent

g.   in an experiment, those objects <u>not</u> exposed to the variable being tested

h.   group of <u>facts</u> from which conclusions may be drawn

i.   an experiment

j.   a collection of related hypotheses

Multiple Choice Questions

1. A "good" hypothesis
   a. should explain unique unobserved phenomena.
   b. should be testable and consistent with all observations collected up to that point.
   c. should be testable, but may not be consistent with all observations collected up to that point.
   d. need not be testable, but must be consistent with all observations collected up to that point.
   e. should be repeatable, but may not be consistent with all observations collected up to that point.

2. An experiment should
   a. make predictions and be repeatable.
   b. test a hypothesis and be consistent with all observations collected up to that point.
   c. make predictions and be consistent with all observations collected up to that point.
   d. test predictions, but need not be consistent with all observations collected up to that point.
   e. test a hypothesis but need not be consistent with all observations collected up to that point.

3. An experiment may attempt to
   a. establish blame.
   b. establish cause and effect.
   c. repeat all observations collected up to that point.
   d. establish cause and effect or repeat all observations collected up to that point.
   e. make predictions.

4. An experiment may attempt to
   a. establish blame.
   b. establish a relationship between all observations collected up to that point.
   c. repeat all observations collected up to that point.
   d. establish cause and effect or repeat all observations collected up to that point.
   e. make predictions.

5. A spaceship from the planet Venus came to earth to take a sample of earthlings back for study. It landed near the Fabulous Forum, home of the Los Angeles Lakers. The visitors quickly captured 10 earthlings from the center of the basketball court and returned to Venus to study their "specimens". Will the data they gather be accurate and appropriate to use in proposing hypotheses about earthlings?
   a. No. Their sample is biased.
   b. No. There is no control group.
   c. Yes. A control group is not required.
   d. Yes. Their sample is not biased.
   e. It is not possible to tell from the description given.

6. Each time my neighbor Sam barks (Sam's a dog) at 5 A.M. for 1/2 hour or more, there is an earthquake somewhere the following day. This has happened 20 times in the last eight years. I have concluded from this that Sam can predict earthquakes. Pick the best statement concerning my conclusion.
   a. Since the sample is unbiased, my conclusions are valid.
   b. I have eliminated all variables except Sam's barking, and my conclusions are supported.
   c. There are inadequate controls and, therefore, no clear correlation between Sam's barking and the occurrence of earthquakes.
   d. Although there is an appropriate control, my sample is biased.
   e. The data do not support my hypothesis.

NOTE: Answers for questions 7 through 12 are provided in the Appendix. Questions 13 through 18 constitute a similar sequence, but the answers are not given.

Use the following information for questions 7-12.

A pharmaceutical company has developed a new drug. In preliminary tests, the drug reduced high blood pressure in laboratory animals when injected under the skin. A second experiment was designed to more thoroughly test the drug's effectiveness. One thousand male rabbits were tested. Five hundred received the drug through injection (Group "A") and five hundred received nothing (Group "B"). All rabbits were the same breed, fed the same type and amount of food, and housed under identical conditions. After the drug was administered, Group "A" had elevated blood pressure for approximately one hour, then it returned to normal. The blood pressure of Group "B" remained unchanged and normal.

7. Which group of rabbits acted as a control?
    a   Group "A"
    b. Group "B"
    c. neither group
    d. all 1000 rabbits
    e. half of group "A" and half of group "B"

8. Were the control and experimental groups adequately matched?
    a. Yes, because only male rabbits were used.
    b. Yes, because they were all the same breed, fed the same type and amount of food, and housed under identical conditions.
    c. No, because no female rabbits were used.
    d. No, because a Group "B" did not receive an injection.
    e. No, because only Group "A" received the drug.

9. Assuming an adequate control was used, which hypothesis could be tested using this experimental design?
    a. The drug lowers blood pressure in some breeds of male rabbits but not in others.
    b. The drug does not lower blood pressure in male rabbits kept in the dark.
    c. The drug lowers blood pressure in male rabbits.
    d. The drug does not lower blood pressure when taken orally.
    e. The drug lowers blood pressure when taken orally.

10. Assuming an adequate control was used, which hypothesis could be tested using this experimental design?
    a. The drug does not lower blood pressure in overweight rabbits.
    b. The drug does not lower blood pressure in male rabbits.
    c. The drug lowers blood pressure in male rabbits housed in cages smaller than 0.5 X 0.5 X 0.5 meters.
    d. The drug lowers blood pressure in Group "B" rabbits
    e. The drug does not lower blood pressure in Group "B" rabbits.

11. Why were the rabbits "...of the same breed, were fed the same type and amount of food, and were housed under identical conditions."?
    a. to simplify their care and maintenance during the experiment
    b. to standardize the test
    c. to reduce possible psychological bias
    d. to reduce the number of variables
    e. to assure the rabbits have a normal heart rate

12. If the control group receives a placebo, what purpose would it serve?
  a. It would simplify their care and maintenance during the experiment.
  b. It would standardize the test.
  c. It would reduce possible psychological bias.
  d. It would reduce the number of variables.
  e. It would assure the use of a placebo.

● Use the following information for questions 13-20.

Melanoma is a type of cancer that produces a black-pigmented skin tumor. Its incidence is on the rise. For this and other reasons, a great deal of research is concentrating on this type of cancer. It has been hypothesized that injections of a vaccine containing chemical substances shed by melanoma cells grown in a research lab might stimulate the body's immune system (its natural defense system), thereby helping to reduce the tumor's growth. To test this hypothesis, one hundred melanoma patients were given the experimental vaccine by injection (Group "A"), and one hundred melanoma patients were given an injection of a placebo(Group "B"). All subjects were matched according to age, gender, and diet, and all reside in the United States. Group "A" subjects were from the midwest and Group "B" from the southeast. After the vaccine was administered, Group "A" showed a reduced rate of tumor growth. Tumor growth in Group "B" remained unchanged.

13. Which was the experimental group?
  a. Group "A"
  b. Group "B"
  c. neither group
  d. all 100 patients
  e. half of group "A" and half of group "B"

14. Were the control and experimental groups adequately matched?
  a. Yes, because the two groups are from different geographic regions.
  b. Yes, because they were matched according to age, gender, and diet.
  c. No, because both groups had melanoma.
  d. No, because Group "B" did not have melanoma.
  e. No, because they were not matched for general life style and geographical area of residence.

15. Assuming an adequate control was used, which hypothesis could be tested using this experimental design?
  a. The vaccine reduces tumor growth in males but not in females.
  b. The vaccine does not reduce tumor growth in subjects who smoke.
  c. The vaccine reduces tumor growth.
  d. The vaccine does not reduce tumor growth when applied to the skin.
  e. The vaccine reduces tumor growth when applied to the skin.

16. Assuming an adequate control was used, which hypothesis could be tested using this experimental design?
  a. The vaccine does not reduce tumor growth in people who smoke.
  b. The vaccine does not reduce tumor growth.
  c. The vaccine reduces tumor growth in people who live in rural areas.
  d. The vaccine reduces tumor growth in subjects in Group "B".
  e. The vaccine reduces tumor growth in Group "B".

17. Why were the subjects "...matched according to age, gender, and diet"?
  a. to simplify their care and maintenance during the experiment
  b. to reduce the number of variables
  c. to reduce possible psychological bias
  d. to standardize the test
  e. to assure the tumors have a normal growth rate

18. The control group received a placebo. What purpose did it serve?
    a. It simplified their care and maintenance during the experiment.
    b. It standardized the test.
    c. It reduced possible psychological bias.
    d. By general agreement, researchers always use a placebo.
    e. It reduced the number of variables.

19. The researchers knew if a subject belonged to the experimental or control group in this study, but the subjects did not. Therefore, the design is called
    a. a blind procedure.
    b. a double-blind procedure.

20. Neither researchers nor subjects knew who belonged to the experimental or control group. Therefore, the design is called
    a. a blind procedure.
    b. a double-blind procedure.

Concept Map Construction

Construct a concept map for each group of terms. Be sure to include appropriate connector phrases. You may add other terms as necessary and use terms in the plural or singular form.

1. theory of evolution, hypothesis, experiment, double-blind, variable
2. *Penicillium*, antibiotic, fungus, bacterial infection, repeatability
3. psychological bias, variable, experimental group, placebo, observation

Chapter 3

The Atomic Basis of Life

Section Concept Map

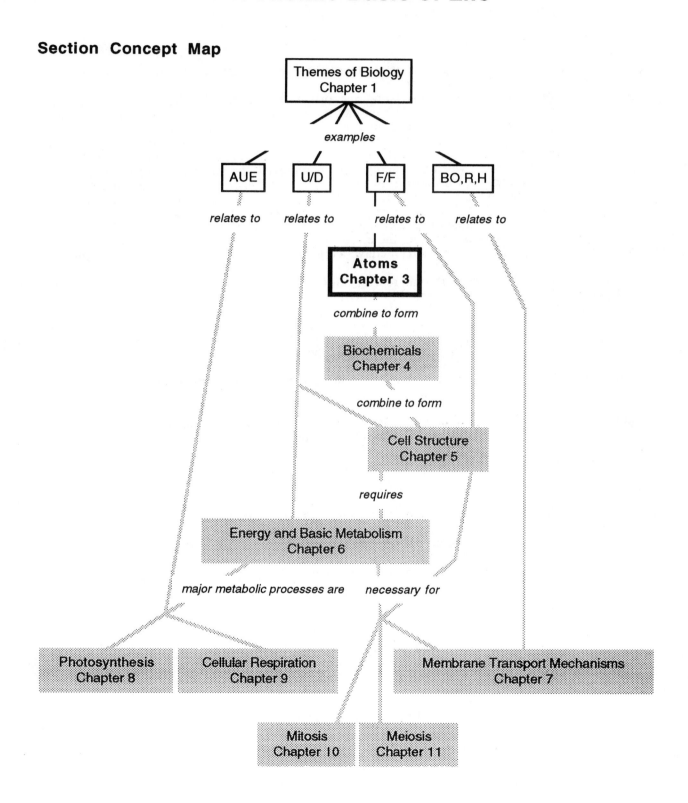

Go Figure!

Figure 3-4:   The number in the center of each atom represents the number of positive charges in the atom. What is this number called?

Where are the positive charges located?

Which subatomic particle carries a positive charge?

Which subatomic particle is *not* represented in these atomic structure diagrams?

Figure 3-5:   Movement of electrons from one orbital to another results in an increase of potential energy of the electron and it takes energy to do this. A similar situation occurs when a ball is rolled up a hill. An energy input is required to move the ball against the pull of gravity. Some of this energy is stored as potential energy. If the ball does not reach a position of stability (*i.e.,* resting on a flat spot or lodged behind a rock) when the energy input is removed, the ball will spontaneously begin rolling back down the hill all on its own. The energy for this movement comes from the stored potential energy until the ball reaches a position of stability. The same applies to the electron. As long as the energy source is being used, the electron can remain in a less stable, higher energy orbital in the atom. But as soon an the energy is removed, the electron drops to a more stable, lower energy orbital. The potential energy is then converted to some other form, such as heat or light.

Figure 3-6:   Notice the two different representations of each molecule. The upper drawing shows atomic nuclei and their electrons. The lower uses elemental symbols.

How many electrons must be shared to get a covalent bond?

What does each line represent between the letters?

Figure 3-8:   What is the maximum number of hydrogen bonds a water molecule can form at any one instant?

Figure 3-9:   Which is a more stable (*i.e.,* less likely to change) arrangement of molecules: those in the upper picture or those in the lower picture?

Figure 3-10: Why are the water molecules oriented differently around the sodium (Na^+) compared to the chloride (Cl^-) ions?

Figure 3-13: Which is more acidic: coffee or acetic acid? By what factor?

Which is more acidic: coffee or baking soda? By what factor?

Which is more acidic: soap or baking soda? By what factor?

Matching Questions

Write the letter of the phrase that best matches the numbered term on the left. Use each only once.

_____ 1. element

_____ 2. electron

_____ 3. isotope

_____ 4. atomic number

_____ 5. orbital

_____ 6. covalent bond

_____ 7. polar covalent bond

_____ 8. nonpolar compound

_____ 9. hydrogen bond

_____10. hydrophilic molecule

_____11. solvent

_____12. solution

_____13. pH

_____14. cohesion of water molecules

_____15. calorie

_____16. acid

_____17. buffer

_____18. acidic solution

_____19. free radical

_____20. water

a. this equals the number of protons in the atom of an element

b. formed by equal sharing of electrons

c. a chemical that resists changes in pH by combining with H^+ or OH^- ions

d. a weak attraction between polar molecules

e. this consists of a solute and a solvent

f. this has an unpaired electron

g. this compound may act as an acid or a base

h. negatively charged subatomic particle

i. formed by an unequal sharing of electrons

j. the location of electrons in an atom

k. a measure of H^+ concentration in a solution

l. a measure of heat energy

m. a substance that readily gives up H^+

n. a fundamental form of matter

o. this has a pH less than 7.0

p. a substance that dissolves another substance

q. the property of water molecules responsible for its high surface tension

r. the molecules have an equal distribution of charges within them

s. it differs in its atomic mass from other atoms of the same element

t. this often contains a polar covalent bond

Multiple Choice Questions

1. Which symbol is used to designate the element carbon?
    a. C
    b. Cl
    c. Ca
    d. Fe
    e. K

2. Which symbol is used to designate the element nitrogen?
    a. H
    b. Na
    c. O
    d. N
    e. Mg

3. Which describes an atom?
    a. It is a solid sphere.
    b. The nucleus contains protons and neutrons.
    c. The number of protons equals the number of neutrons.
    d. The nucleus contains electrons.
    e. Protons and neutrons have the same charge.

4. Which describes an atom?
    a. Most of the atom is empty space.
    b. Protons and electrons have the same charge.
    c. Electrons are positively charged.
    d. Neutrons and electrons have the same mass.
    e. Protons and electrons have the same mass.

5. Which is a false statement?
    a. Atomic mass is sometimes referred to as "atomic weight".
    b. Atomic mass may differ among atoms of the same element.
    c. Electrons, neutrons, and protons contribute significantly to atomic mass.
    d. Isotopes of an element have a different number of neutrons.
    e. Radioactive isotopes are unstable and break apart spontaneously.

6. Which is a false statement?
    a. All atoms of the same element have the same atomic number.
    b. All atoms of the same element have the same number of protons.
    c. Electrons occupying shells closer to the nucleus have more energy than those farther away from the nucleus.
    d. Electron orbitals may contain a maximum of two electrons.
    e. A neutrally charged atom will have an equal number of electrons and protons.

7. The atomic number of phosphorus is 15. How many electrons are in each shell of an atom of phosphorus? (Begin with the number in the innermost shell.)
    a. 15:0
    b. 5:8:2
    c. 2:6:7
    d. 2:8:5
    e. 2:2:2:2:2:2:2:1

8. The atomic number of magnesium is 12. How many electrons are in each shell of an atom of magnesium? (Begin with the number in the innermost shell.)
    a. 2:2:8
    b. 2:4:6
    c. 6:4:2
    d. 2:2:2:2:2:2
    e. 2:8:2

9. Which DOES NOT describe covalent bonds?
    a. Electrons are shared between atoms.
    b. The resulting structure is referred to as an ion.
    c. Electrons travel around the nuclei of both atoms.
    d. A double bond involves two pair of electrons.
    e. Some atoms can form more than one covalent bond.

10. Which DOES NOT describe ionic bonds?
    a. Ionic bonds form between charged atoms.
    b. Ionic bonds require complete transfer of an electron from one atom to another.
    c. Some atoms can form more than one ionic bond.
    d. Ionic bonds always involve pairs of electrons.
    e. Ionic bonds are one example of noncovalent bonds.

11. How many electrons must be gained by an atom of sulfur (atomic number = 16) in order to fill the outer electron shell?
    a. 1
    b. 2
    c. 3
    d. 4
    e. 5

12. How many electrons must be gained by an atom of silicon (atomic number = 14) in order to fill the outer electron shell?
    a. 1
    b. 2
    c. 3
    d. 4
    e. 5

13. What is the maximum number of covalent bonds an atom of phosphorus (atomic number = 15) can form?
    a. 1
    b. 2
    c. 3
    d. 4
    e. 5

14. What is the maximum number of covalent bonds an atom of silicon (atomic number = 14) can form?
    a. 1
    b. 2
    c. 3
    d. 4
    e. 5

15. How many electrons must an atom of magnesium (atomic number = 12) lose from its outer shell in order to maximize its stability?
   a. 1
   b. 2
   c. 3
   d. 4
   e. 5

16. How many electrons must an atom of fluorine (atomic number = 9) gain in its outer shell in order to maximize its stability?
   a. 1
   b. 2
   c. 3
   d. 4
   e. 5

17. Which DOES NOT describe polar molecules?
   a. They have regions of electrical charge within them.
   b. They result from unequal sharing of electrons in a covalent bond.
   c. They often participate in hydrogen bonds.
   d. The number of electrons exceeds the number of protons in the molecule.
   e. They are hydrophilic.

18. Which describes nonpolar molecules?
   a. They are ions.
   b. They are hydrophobic.
   c. They have regions of electrical charge within them.
   d. The number of electrons exceeds the number of protons in the molecule.
   e. They often participate in hydrogen bonds.

19. Which describes an acid?
   a. It is a substance that lowers the pH of a solution.
   b. If it is a strong acid, it will not give up its proton very easily.
   c. It is a substance that accepts protons from water.
   d. It will not react with a buffer, but may react with water.
   e. It will not react with OH^- ions, but may react with a buffer.

20. Which describes a base?
   a. It is a substance that lowers the pH of a solution.
   b. It binds readily to H^+ ions.
   c. It binds readily to OH^- ions.
   d. Water never acts as a base.
   e. Addition of a base would tend to make a solution less alkaline.

Concept Map Construction

Construct a concept map for each group of terms. Be sure to include appropriate connector phrases. You may add other terms as necessary and use terms in the singular or plural form.

   1. scientific method, element, sulfur, subatomic particle, radioisotope
   2. homeostasis, electron orbital, covalent bond, hydrophobic compound, solvent
   3. hypothesis, surface tension, hydrogen bond, nonpolar molecule, neutron

Chapter 4

Biochemicals: The Molecules of Life

Section Concept Map

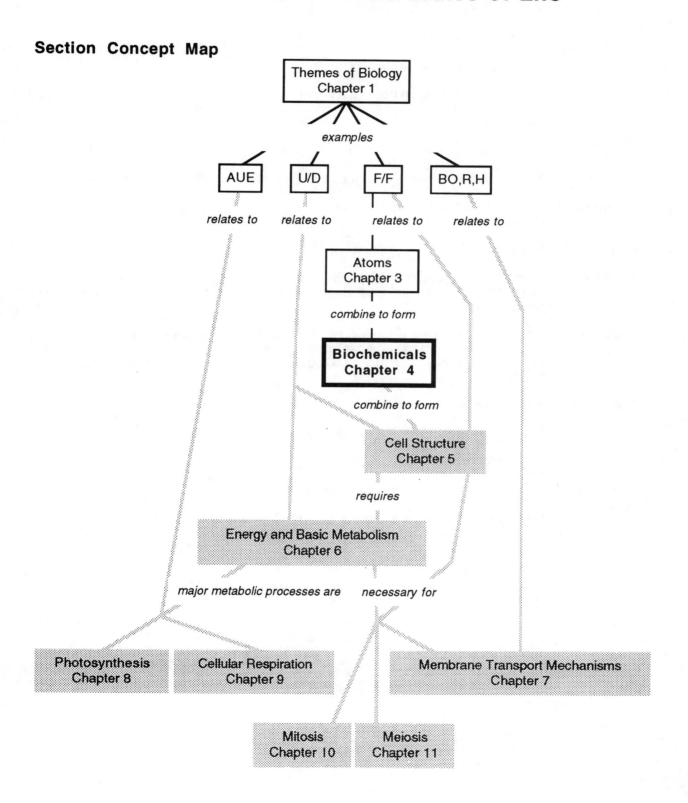

Go Figure!

Figure 4-3:   What happens to the "carrier" after it has delivered the monomer to the growing polymer?

Figure 4-4:   What is the chemical formula for fructose?

What is the chemical formula for ribose?

Are fructose and ribose isomers?

Notice that fructose and ribose are represented by a drawing intermediate between styles (1) and (3) used for glucose since some, but not all, of the carbons are shown. Draw fructose and ribose using style (3).

Why do you suppose chemists use different styles?

Figure 4-9:   What difference in the third fatty acid chain might be responsible for its bend?

Figure 4-11:   Use the functional groups in Figure 4-2 to describe how testosterone and estrogen differ.

Figure 4-13:   What kind of bond is a peptide bond?

Figure 4-14:   Compare the representations of amino acids shown here with those in Figure 4-13. Might the altered positions of amino and R groups affect the shape or function of the molecule?

Figure 4-15:   Notice that each level of organization in a protein retains lower ones. The primary, secondary, and tertiary structure of the polypeptide subunits is still visible in the drawings of secondary, tertiary, and quaternary structure, respectively. Find each of these.

Figure 4-16:   What is the unfolding called?

What levels of structure are lost during unfolding?

What level of structure is unaffected?

Figure 4-19:   What kind of bond is the sugar-phosphate linkage?

How many nucleotides are shown in this polynucleotide?

Matching Questions

Write the letter of the phrase that best matches the numbered term on the left. Use each only once.

_____ 1.   organic molecule

_____ 2.   steroid

_____ 3.   triglyceride

_____ 4.   polymer

_____ 5.   polyunsaturated fat

_____ 6.   carbohydrate

_____ 7.   condensation

_____ 8.   R group

_____ 9.   functional group

_____10.   hydrolysis

_____11.   deoxyribonucleic acid (DNA)

_____12.   peptide bond

_____13.   alpha helix

_____14.   quaternary structure

_____15.   adenosine triphosphate (ATP)

_____16.   nucleotide

_____17.   monosaccharide

_____18.   amino acid

_____19.   denaturation

_____20.   conformational change

a.   a molecule with many double bonds in it

b.   reaction used during macromolecule synthesis

c.   a chemical reaction in which water splits a bond

d.   a nucleotide used as an energy source

e.   the molecule of heredity

f.   the basic molecule has a molecular skeleton of four carbon rings

g.   normal alteration of tertiary structure

h.   the family of sugar molecules

i.   the subunit of protein

j.   a chemical bond between amino acids

k.   a general term applied to a molecule that contains carbon

l.   arrangement of polypeptide subunits in a protein

m.   the subunit of nucleic acids

n.   a simple sugar

o.   a fat molecule

p.   loss of tertiary structure by a protein

q.   one type of secondary structure in proteins

r.   this makes amino acids different

s.   this gives organic molecules special properties

t.   a large molecule made up of many subunits

Multiple Choice Questions

1. Which is an organic molecule?
   a. $C_3H_8O_3$
   b. NO_2
   c. H_2O
   d. H_2SO_4
   e. O_2

2. Which is an organic molecule?
   a. HCl
   b. H_2S
   c. H_2
   d. $CH_3COCOOH$
   e. NaCl

3. Which DOES NOT describe organic molecules?
   a. Organic molecules contain the element carbon.
   b. Organic molecules may be linear, branched, or cyclic.
   c. Organic molecules must have come from a living organism.
   d. Organic molecules may contain functional groups that give them special properties.
   e. Organic molecules may be polar or nonpolar.

4. Which DOES NOT describe biochemicals?
   a. The carbon chain may be linear, branched, or cyclic.
   b. They must have come from a living organism.
   c. They may contain functional groups that give them special properties.
   d. They may be polar or nonpolar.
   e. They are usually simple, small molecules.

5. Which DOES NOT describe biological macromolecules?
   a. They are often very small in size.
   b. They are often polymers.
   c. Their synthesis results from condensation reactions.
   d. Carrier molecules are removed from monomers prior to assembly of certain macromolecules.
   e. They are organic molecules.

6. Which DOES NOT describe biological macromolecules?
   a. Hydrolysis produces monomers from polymers.
   b. There are four basic families of of biological macromolecules.
   c. Macromolecules from different organisms are very different in structure.
   d. Biochemical diversity occurs at the level of monomer sequences.
   e. Each biochemical family has basic functions.

7. Which is a true statement about carbohydrates?
   a. Glycogen is a structural polysaccharide in animals.
   b. Starch is a polysaccharide used for energy storage in plants.
   c. Chitin is a structural polysaccharide in most plants.
   d. Cellulose is easily digested by most animals.
   e. A peptide bond joins two monosaccharides into a disaccharide.

8. Which is a true statement about carbohydrates?
   a. Carbohydrates function as energy storage and structural molecules.
   b. Carbohydrates are typically hydrophobic.
   c. Carbohydrates contain more energy than an equivalent weight of fat.
   d. Monosaccharides function as hormones in many organisms.
   e. Triglycerides are the subunits of polysaccharides.

9. Which molecule is used for short term energy storage?
   a. fat
   b. chitin
   c. glycerol
   d. glycogen
   e. cellulose

10. Which molecule is used for long term energy storage?
    a. fat
    b. chitin
    c. glycerol
    d. glycogen
    e. cellulose

11. Which describes lipids?
    a. Lipids are soluble in water.
    b. Lipids are hydrophilic.
    c. Lipids are nonpolar.
    d. Lipids are inorganic molecules.
    e. Lipids include molecules like polypeptides.

12. Which is a true statement?
    a. Fats contain a phosphate group.
    b. Steroids are made of glycerol and three fatty acids.
    c. Phospholipids resemble steroids in molecular structure.
    d. Unsaturated fats tend to be liquid at room temperature.
    e. Fats have a hydrophilic and a hydrophobic portion.

13. Which class of molecule is the major component of cell membranes?
    a. phospholipid
    b. cellulose
    c. wax
    d. glycogen
    e. triglyceride

14. Which molecule is used to make plant cell walls?
    a. disaccharide
    b. polypeptide
    c. cellulose
    d. phospholipid
    e. chitin

15. Which is a true statement about proteins?
    a. Each organism has relatively few proteins.
    b. Proteins are the primary form of energy storage in most organisms.
    c. Proteins are fatty acid polymers.
    d. Proteins from all organisms contain the same 20 amino acids.
    e. All proteins have the same three dimensional structure.

16. Which DOES NOT describe proteins?
    a. The sequence of amino acids determines the primary structure of a protein.
    b. The secondary structure of a protein may be in the form of a helix, pleated sheet, or random coil.
    c. Physical properties of the R groups in the protein help determine tertiary structure.
    d. Denaturation of a protein involves loss of primary structure.
    e. Flexible proteins are capable of conformational changes.

17. Which molecule is responsible for storing hereditary information?
    a. deoxyribonucleic acid (DNA)
    b. steroids
    c. ribonucleic acid (RNA)
    d. triglycerides
    e. polypeptides

18. Enzymes are
    a. lipids.
    b. proteins.
    c. carbohydrates.
    d. nucleic acids.
    e. inorganic molecules.

19. What are the components of an RNA nucleotide?
    a. deoxyribose sugar
    b. deoxyribose sugar and a phosphate
    c. deoxyribose sugar, a phosphate, and a nitrogenous base
    d. ribose sugar and a phosphate
    e. ribose sugar, a phosphate, and a nitrogenous base

20. In what way do RNA nucleotides differ?
    a. They have different sugars.
    b. They have different phosphates.
    c. They have different nitrogenous bases.
    d. They have different bonds holding them together.
    e. There are no differences between RNA nucleotides.

Concept Map Construction

Construct a concept map for each group of terms. Be sure to include appropriate connector phrases. You may add other terms as necessary and use terms in the singular or plural form.

1. R group, biochemical, atomic nucleus, ATP, nitrogenous base
2. functional group. amino acid, polymer, lipid, hydrogen bond
3. peptide bond, triglyceride, saturated fat, covalent bond, tertiary structure

Chapter 5

Cell Structure and Function

Section Concept Map

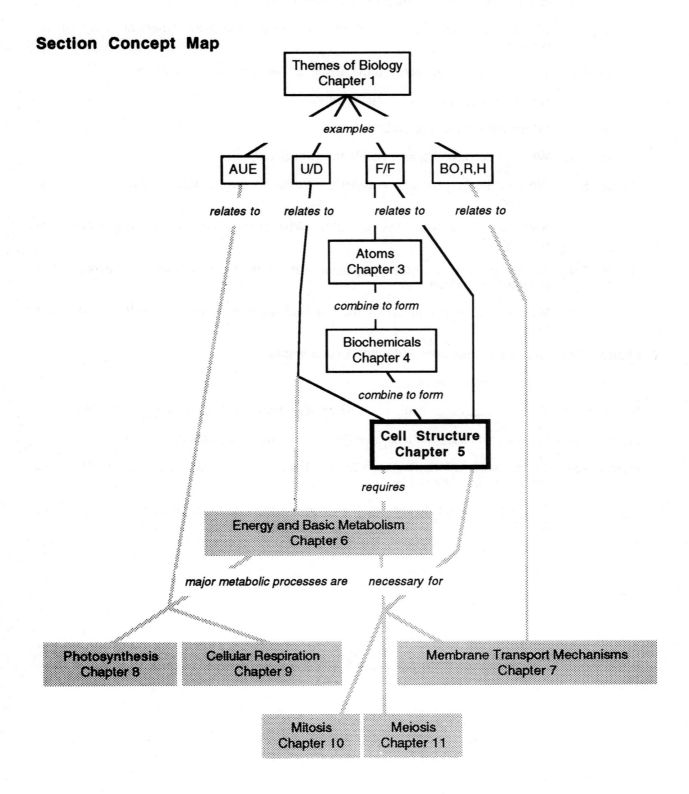

27

Go Figure!

Figure 5-2:   Which differences between prokaryotic and eukaryotic cells are visible in this figure?

Figure 5-4:   What do the globular objects in this model represent?

Why do some go completely through the membrane while others are only on the surface?

Figure 5-5:   Which organelles depicted in this figure are present only in plant cells?

Which are present only in animal cells?

Which are common to both?

Figure 5-6:   Why are no chromosomes visible in the nucleus?

Figure 5-7:   What is the purpose of the genetic material becoming tightly coiled prior to cell division?

Notice the chromosome has two parts. What process has occurred to produce those two parts?

Figure 5-9:   Are the ribosomes on the cytoplasmic side or the internal side of the endoplasmic reticulum membranes?

Figure 5-10: Why are Golgi complexes "clustered near the area where new cell wall is being manufactured"?

Figure 5-15: How are chloroplasts and mitochondria similar?

How are they different?

Figure 5-13: Which organelle is involved in the cellular death resulting in finger formation?

Figure 5-18: What could cause the leading edge of a moving cell to contain "ruffles?"

Figure 5-20: In what ways do the flagellum in this figure and the cilia in Figure 5-19 differ? In what ways are they similar?

Figure 5-21: Specifically, what do the numbers 9 and 2 refer to in the phrase "9 + 2 array of microtubules"?

Matching Questions

Write the letter of the phrase that best matches the numbered term on the left. Use each only once.

_____ 1. simple microscope

_____ 2. exocytosis

_____ 3. microtubule

_____ 4. smooth endoplasmic reticulum

_____ 5. prokaryotic cell

_____ 6. crista

_____ 7. Golgi complex

_____ 8. microfilament

_____ 9. cytoplasm

_____10. lysosome

_____11. cilium

_____12. phospholipid bilayer

_____13. primary cell wall

_____14. thylakoid

_____15. vacuole

_____16. ribosome

_____17. receptor

_____18. nuclear envelope

_____19. communicating junction

_____20. nucleolus

a. composed of the protein tubulin

b. composed of the protein actin

c. composed of cellulose

d. produces secretory vesicles

e. a primitive type of cell

f. formed by the chloroplast inner membrane

g. formed by the mitochondrial inner membrane

h. uses a single lens to magnify

i. especially prominent in mature plant cells

j. the basic structure of membrane

k. membrane protein involved in recognition of substances

l. the final step in the secretory pathway

m. allows movement of materials between cells

n. site of protein synthesis

o. consists of a double membrane

p. a short projection used for movement

q. site of ribosome formation

r. the cellular region outside the nucleus

s. contains hydrolytic enzymes

t. involved in the synthesis of steroid hormones

Multiple Choice Questions

1. Which is a true statement?
   a. Cells were discovered by Anton van Leeuwenhoek in the 1600s.
   b. Cells were first seen in pond water.
   c. The discovery of cells required the invention of the light microscope.
   d. The name "cell" was first used by Theodor Schwann.
   e. Cells are found only in plants.

2. Which DOES NOT describe the cell theory?
   a. All living things are made of cells.
   b. Cells may develop from nonliving matter.
   c. Cells are the smallest organizational unit that have all the properties of life.
   d. Theodor Schwann and Matthias Schleiden contributed to development of the cell theory.
   e. Rudolf Virchow contributed to the development of the cell theory.

3. Which is a true statement concerning prokaryotic cells?
   a. Prokaryotic cells are generally larger than 100μm.
   b. Prokaryotic cells have a well-defined nucleus.
   c. Prokaryotic cells are found in bacteria.
   d. Prokaryotic cells evolved from eukaryotic cells.
   e. Prokaryotic cells lack ribosomes.

4. Which is a true statement concerning eukaryotic cells?
   a. Eukaryotic cells are capable of more complex activities than prokaryotic cells.
   b. Eukaryotic cells are usually smaller than 10μm.
   c. Eukaryotic cells lack membranous organelles.
   d. Eukaryotic cells lack ribosomes.
   e. Eukaryotic cells are simpler in structure than prokaryotic cells.

5. Which IS NOT a component of animal plasma membranes?
   a. phospholipid
   b. carbohydrate
   c. nucleic acid
   d. protein
   e. cholesterol

6. Glycoproteins protruding from the surface of a membrane may act as
   a. carrier molecules.
   b. receptors.
   c. hormones.
   d. antibodies.
   e. cilia.

7. Which DOES NOT describe the nucleus of eukaryotic cells?
   a. It contains most of the DNA of the cell.
   b. The nuclear envelope consists of a single membrane.
   c. Chromosomes are visible only when a cell is dividing.
   d. Chromosome numbers differ between species.
   e. Nuclear pores in the envelope allow passage of materials between the nucleus and the cytoplasm.

8. Which DOES NOT describe endoplasmic reticulum?
   a. Rough endoplasmic reticulum is associated with ribosomes.
   b. Rough endoplasmic reticulum is most abundant in secretory cells.
   c. Rough endoplasmic reticulum is part of the secretory pathway.
   d. Smooth endoplasmic reticulum may destroy toxic substances.
   e. Smooth endoplasmic reticulum is involved in protein synthesis.

9. Which organelle IS NOT made of membrane?
   a. endoplasmic reticulum
   b. chloroplast
   c. Golgi complex
   d. lysosome
   e. ribosome

10. Which IS NOT made of protein?
    a. microtubule
    b. microfilament
    c. intermediate filament
    d. vesicle
    e. cytoskeleton

11. Which is the proper sequence of the secretory pathway?
    a. endoplasmic reticulum, Golgi complex, ribosome
    b. Golgi complex, endoplasmic reticulum, chloroplast
    c. mitochondrion, vesicle, cilia
    d. endoplasmic reticulum, Golgi complex, vesicle
    e. Golgi complex, endoplasmic reticulum, mitochondrion

12. Which process is involved in secretion?
    a. exocytosis
    b. hydrolysis
    c. endocytosis
    d. ciliary action
    e. phagocytosis

13. Which DOES NOT describe lysosomes?
    a. They are products of the Golgi complex.
    b. They contain hydrolytic enzymes.
    c. They are active during some developmental processes.
    d. They may be responsible for certain diseases.
    e. They are the site of aerobic respiration.

14. Which DOES NOT describe vacuoles?
    a. They are prominent in many mature plant cells.
    b. They are the site of photosynthesis.
    c. They are made of a single membrane.
    d. They help maintain internal water pressure.
    e. They accumulate wastes from the cell.

15. Which DOES NOT describe chloroplasts?
    a. They perform photosynthesis.
    b. They contain DNA and ribosomes.
    c. They are made of a double membrane.
    d. They are found in plants and fungi.
    e. Thylakoid membranes contain chlorophyll.

16. Which DOES NOT describe mitochondria?
    a. They are made of a double membrane.
    b. They contain DNA and ribosomes.
    c. They are most abundant in inactive cells.
    d. They are found in plants and fungi.
    e. They are involved in ATP production.

17. Which describes microfilaments?
    a. They are rigid strands lipid.
    b. They are involved in cell movement.
    c. They are hollow tubules made of the protein tubulin.
    d. They are evenly distributed throughout the cytoplasm.
    e. They are flexible polymers of carbohydrate.

18. Which describes flagella?
    a. They are shorter than cilia.
    b. They move due to contraction of actin microfilaments.
    c. They contain variable numbers of microtubules.
    d. They propel cells by moving in a whiplike fashion.
    e. They are more abundant than cilia.

19. Which describes plant cell walls?
    a. They are made of the protein cellulose.
    b. Secondary walls are found closer to the cell membrane than primary walls.
    c. Lignin is commonly found in primary walls to make them more flexible.
    d. They prevent water from entering cells.
    e. The primary wall restricts growth of the cell.

20. Which describes occluding junctions?
    a. They are common in plants.
    b. They are formed by a fusion of the walls of adjacent cells.
    c. They restrict movement of materials between cells.
    d. They contain adhesive material between cells.
    e. They form membranous channels between adjacent cells.

21. Which describes the endosymbiosis hypothesis?
    a. It is used to explain the origin of all membranous organelles.
    b. It is used to explain the origin of ribosomes.
    c. It is used to explain the origin of prokaryotic cells.
    d. It is used to explain the origin of chloroplasts.
    e. It is used to explain the origin of Golgi complexes.

22. Which IS NOT evidence supporting the endosymbiosis hypothesis?
    a. Mitochondria have their own DNA similar to prokaryotic DNA.
    b. Chloroplasts have their own ribosomes similar to prokaryotic ribosomes.
    c. Endosymbiotic relationships have been observed in modern prokaryotes.
    d. The inner membrane of mitochondria resembles the plasma membrane of respiratory bacteria.
    e. The outer membrane of chloroplasts resembles the plasma membrane of photosynthetic bacteria.

Concept Map Construction

Construct a concept map for each group of terms. Be sure to include appropriate connector phrases. You may add other terms as necessary and use terms in the singular or plural form.

1. rough endoplasmic reticulum, protein, phospholipid, microfilament, primary cell wall
2. Golgi complex, carbon, hydrolysis, mitochondrion, endosymbiotic hypothesis
3. flagellum, chromosome, DNA, steroid, eukaryote

Chapter 6

Energy, Enzymes and Metabolic Pathways

Section Concept Map

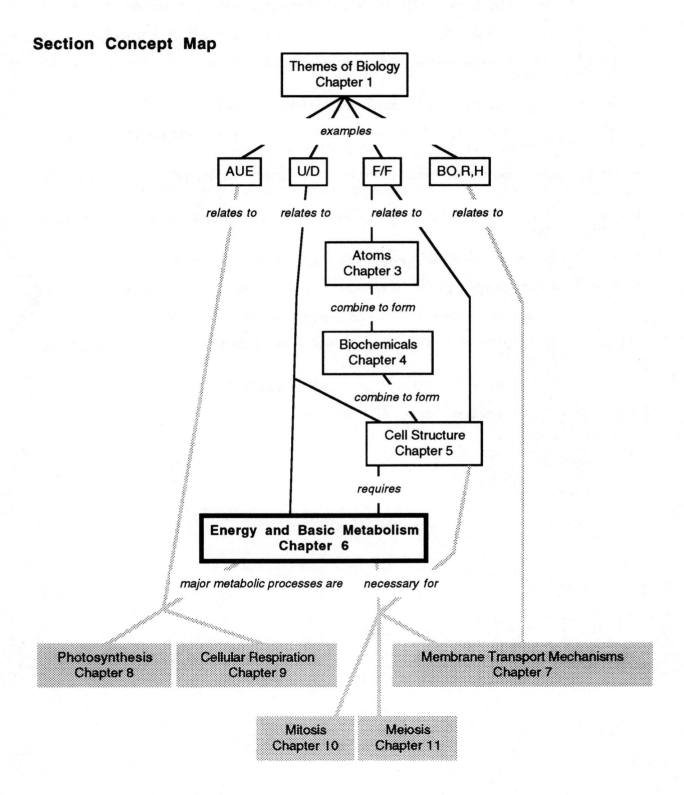

Go Figure!

Figure 6-1:   Is the metabolic pathway in which the frog synthesizes blue pigment likely to be anabolic or catabolic?

Are the reactions in which the Venus fly trap digests the frog anabolic or catabolic?

How about the reactions the Venus fly trap uses to build new plant tissue?

Figure 6-2:   Which laws of thermodynamics are illustrated by the energy transfers shown in this figure?

Figure 6-3:   How could this house be restored to "prime real estate?"

Would restoration of the house violate the second law of thermodynamics since it represents an increase in order? Justify your answer.

Figure 6-5:   Could the kinetic energy of the object moving to ring the bell in this figure move the mallet as much as it was swung in the first place? Why or why not?

Figure 6-6:   Which reaction (exergonic or endergonic) could represent your lunch being digested?

Which reaction could represent new muscle tissue being made?

Figure 6-7:   Do the energy levels of the reactants and products differ with or without enzyme?

Is the reaction illustrated exergonic or endergonic?

Figure 6-8:   Is the conversion of Substrates A and B to Product C in Figure b catabolic or anabolic?

How about the conversion of Substrate C to A and B?

Figure 6-12:   What will activate Enzyme BC if it has been inhibited by Product E?

Matching Questions

Write the letter of the phrase that best matches the numbered term on the left. Use each only once.

_____ 1.   potential energy

_____ 2.   kinetic energy

_____ 3.   cofactor

_____ 4.   endergonic reaction

_____ 5.   exergonic reaction

_____ 6.   activation energy

_____ 7.   substrate

_____ 8.   first law of thermodynamics

_____ 9.   catabolic pathway

_____10.   anabolic pathway

_____11.   oxidation reaction

_____12.   reduction reaction

_____13.   ATP

_____14.   active site

_____15.   NADPH

_____16.   entropy

_____17.   feedback inhibition

_____18.   enzyme

_____19.   protein kinase

_____20.   coupled reactions

a.   an inorganic chemical that assists an enzyme

b.   this binds specifically to an enzyme

c.   an increase in product decreases its own production

d.   a reaction in which energy is released

e.   this reaction involves gaining an electron

f.   the stored form of energy

g.   an enzyme involved in regulating other enzymes

h.   a reaction in which the products have more energy than the reactants

i.   ATP hydrolysis and an endergonic reaction, for example

j.   the location of substrate binding

k.   a measure of disorder

l.   an organic catalyst

m.   the active form of energy

n.   energy required to start a chemical reaction

o.   "energy can neither be created nor destroyed"

p.   series of reactions in which simple molecules are made into a complex one

q.   reserves of this constitute "reducing power"

r.   a nucleotide which supplies energy to chemical reactions

s.   this reaction involves the loss of an electron

t.   series of reactions in which a complex molecule is made into simpler ones

Multiple Choice Questions

1. Which is an example of kinetic energy?
   a. an ionic bond
   b. an automobile battery
   c. ATP
   d. light
   e. glycogen

2. Which is an example of potential energy?
   a. an electric current
   b. wind
   c. a covalent bond
   d. a flowing river
   e. heat

3. "The amount of energy in the universe is finite" is another way of stating the
   a. cell theory.
   b. endosymbiotic hypothesis.
   c. cell invagination hypothesis.
   d. first law of thermodynamics.
   e. second law of thermodynamics.

4. Which IS NOT consistent with the second law of thermodynamics?
   a. Entropy increases.
   b. The amount of usable energy decreases.
   c. Organisms use energy to maintain their organization.
   d. The chemical energy of food is used over and over as an organism lives it life.
   e. Events tend to occur from a higher to a lower energy state.

5. Which describes ATP?
   a. Its hydrolysis is an endergonic reaction.
   b. Its sugar is deoxyribose.
   c. Each molecule contains only two phosphates.
   d. Its hydrolysis is energetically favored in the cell.
   e. Its hydrolysis decreases entropy.

6. Which describes ATP?
   a. Its hydrolysis can be coupled to endergonic reactions of metabolism.
   b. It is a chemical used for long term energy storage.
   c. The amount of energy released during its hydrolysis is equal to or less than the amount of energy stored in a coupled endergonic reaction.
   d. One product of its hydrolysis is water.
   e. Cells maintain large concentrations of it.

7. Which describes enzymes?
   a. They are carbohydrates.
   b. They catalyze chemical reactions.
   c. Each type reacts with a variety of substrates.
   d. Each type is found in large concentrations in the cell.
   e. They increase the activation energy of a reaction.

8. Which describes enzymes?
   a. They bind specific substrates at their active site.
   b. Each may be used only once since they are chemically changed at the end of a reaction.
   c. All enzymes have the same optimum pH.
   d. Since enzymes are chemicals, an increase in temperature always increases the rate of reaction.
   e. Coenzymes are inorganic molecules that assist enzymes in their function.

9. Which DOES NOT describe a catabolic pathway?
    a. Energy is released.
    b. The pathway is exergonic.
    c. It is a biosynthetic pathway.
    d. The reactants are more complex than the products.
    e. All the steps are catalyzed by different enzymes.

10. Which DOES NOT describe an anabolic pathway?
    a. The products are more complex than the reactants.
    b. Although energy is stored, entropy increases.
    c. Some intermediate reactions involve ATP hydrolysis.
    d. It never involves electron transfer.
    e. Each metabolic intermediate is the substrate for a different enzyme.

11. Which describes the situation when an atom becomes oxidized?
    a. The atom has gained an electron.
    b. The atom has gained energy.
    c. The atom may have combined with a hydrogen atom.
    d. The atom may have combined with an oxygen atom.
    e. As the atom becomes oxidized, it is not reacting with any other atom.

12. Which describes the situation when an atom becomes reduced?
    a. The atom becomes smaller.
    b. The atom may have combined with a hydrogen atom.
    c. The atom may have combined with an oxygen atom.
    d. As the atom becomes reduced, it is not reacting with any other atom.
    e. The oxidation state of the atom is unchanged.

13. How do phosphorylase enzymes regulate enzyme activity?
    a. They add a phosphate group to a coenzyme to activate it.
    b. They remove a phosphate group from a cofactor to activate it.
    c. They add a phosphate group to the substrate so it doesn't fit the active site of the enzyme.
    d. They add a phosphate group to the enzyme so the active site won't bind the substrate.
    e. They remove a phosphate group from a coenzyme to inactivate it.

14. How does feedback inhibition regulate enzyme activity?
    a. The final product of a metabolic pathway activates the first enzyme in the pathway by binding to its active site.
    b. The final product of a metabolic pathway inactivates the first enzyme in the pathway by binding to its active site.
    c. The final product of a metabolic pathway activates the first enzyme in the pathway by binding to its feedback site.
    d. The final product of a metabolic pathway inactivates the first enzyme in the pathway by binding to its feedback site.
    e. The final product of a metabolic pathway inactivates the last enzyme in the pathway by binding to its feedback site.

15. Observe the reaction below.

$$\text{Pyruvate} + CO_2 + \text{ATP} + H_2O \longrightarrow \text{Oxaloacetate} + \text{ADP} + P_i$$

With respect to the conversion of pyruvate and CO_2 to oxaloacetate, which is a true statement?
    a. The reaction is anabolic.
    b. The reaction is catabolic.

16. Observe the reaction below.

$$\text{Lactose} \longrightarrow \text{Galactose} + \text{Glucose}$$

Which is a true statement?
a. The reaction is anabolic.
b. The reaction is catabolic.

17. Observe the reaction below.

$$\text{Pyruvate} + CO_2 + \text{ATP} + H_2O \longrightarrow \text{Oxaloacetate} + \text{ADP} + P_i$$

With respect to the conversion of pyruvate and CO_2 to oxaloacetate, which is a true statement?
a. The reaction is endergonic.
b. The reaction is exergonic.

18. Observe the reaction below.

$$\text{1,3-biphosphoglycerate} + \text{ADP} \longrightarrow \text{3-phosphoglycerate} + \text{ATP}$$

With respect to the conversion of 1,3-biphosphoglycerate to 3-phosphoglycerate, which is a true statement?
a. The reaction is endergonic.
b. The reaction is exergonic.

19. The coenzyme NAD (nicotinamide adenine dinucleotide) is an electron carrier similar to NADP. With that in mind, observe the reaction below.

$$\text{Glyceraldehyde-3-phosphate} + P_i + NAD^+ \longrightarrow \text{1,3-bisphosphoglycerate} + \text{NADH} + H^+$$

Which is a true statement?
a. Glyceraldehyde-3-phosphate is oxidized to 1,3-bisphosphoglycerate.
b. Glyceraldehyde-3-phosphate is reduced to 1,3-bisphosphoglycerate.

20. The coenzyme NAD (nicotinamide adenine dinucleotide) is an electron carrier similar to NADP. With that in mind, observe the reaction below.

$$\text{Pyruvate} + \text{NADH} \longrightarrow \text{Lactate} + NAD^+$$

Which is a true statement?
a. Pyruvate becomes oxidized to lactate.
b. Pyruvate becomes reduced to lactate.

21. Observe the reaction below.

$$\text{Pyruvate} + NH_4^+ + \text{NADH} + H^+ \longrightarrow \text{L-alanine} + NAD^+ + H_2O$$

With respect to pyruvate being converted to L-alanine, which is a true statement?
a. Pyruvate is oxidized to L-alanine and the reaction is anabolic.
b. Pyruvate is reduced to L-alanine and the reaction is anabolic.
c. Pyruvate is oxidized to L-alanine and the reaction is catabolic.
d. Pyruvate is reduced to L-alanine and the reaction is catabolic.

22. Observe the reaction below.

$$\text{Glutamate} + NADP^+ + H_2O \longrightarrow \alpha\text{-ketoglutarate} + NH_4^+ + NADPH + H^+$$

With respect to the conversion of glutamate to L-alanine, which is a true statement?
a. Glutamate is oxidized to a-ketoglutarate and the reaction is anabolic.
b. Glutamate is reduced to a-ketoglutarate and the reaction is anabolic.
c. Glutamate is oxidized to a-ketoglutarate and the reaction is catabolic.
d. Glutamate is reduced to a-ketoglutarate and the reaction is catabolic.

Concept Map Construction

Construct a concept map for each group of terms. Be sure to include appropriate connector phrases.
You may add other terms as necessary and use terms in the singular or plural form.

1. kinetic energy, coenzyme, oxidation reaction, atomic nucleus, carbon
2. NAD, first law of thermodynamics, ribosome, mitochondrion, conformational change
3. feedback inhibition, exergonic reaction, coupled reactions, active site, carbohydrate

Chapter 7

Movement of Materials Across Membranes

Section Concept Map

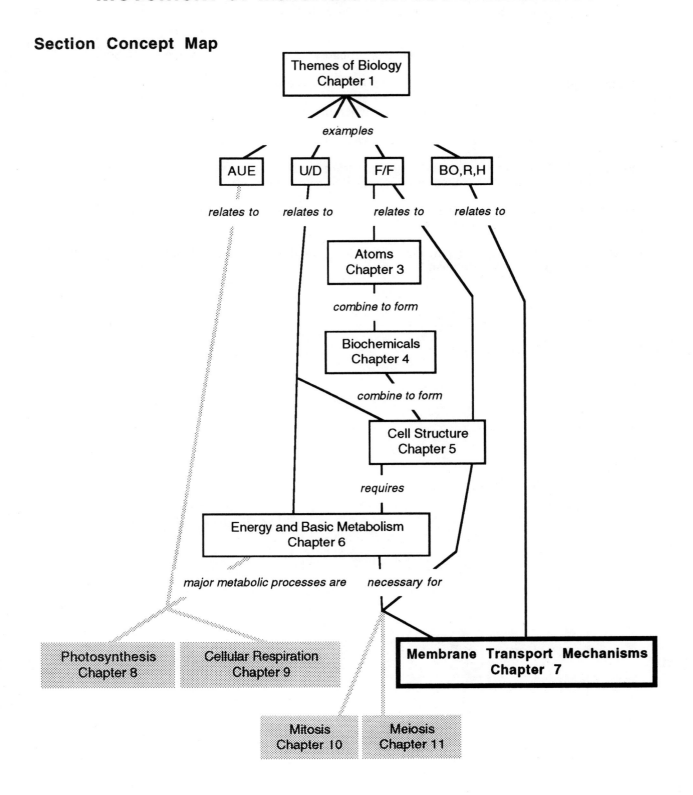

Go Figure!

Figure 7-2: How would the results of urea movement change if the membrane was impermeable to urea?

How would the results of urea movement change if urea was actively transported into the cell?

Figure 7-3: Why are membrane channels probably unnecessary for nonionic compounds?

Figure 7-4: Does any water enter or leave the cell when in an isotonic solution?

Does any water enter the cell when in a hypertonic solution?

Does any water leave the cell when in a hypotonic solution?

Figure 7-5: Why are the terms "turgor pressure" and "plasmolysis" not used in reference to animal cells placed in hypotonic and hypertonic solutions respectively?

Figure 7-9: The enlargement represents an epithelial cell. Why do you suppose its membrane facing the lumen is wavy?

Figure 7-10: Why must the particle in the food vacuole be digested before it can enter the amoeba's cytoplasm?

Matching Questions

Write the letter of the phrase that best matches the numbered term on the left. Use each only once.

_____ 1.   selective permeability

_____ 2.   carrier protein

_____ 3.   turgor pressure

_____ 4.   osmosis

_____ 5.   phagocytosis

_____ 6.   pinocytosis

_____ 7.   simple diffusion

_____ 8.   facilitated diffusion

_____ 9.   active transport

_____10.   ATPase

_____11.   isotonic

_____12.   hypertonic

_____13.   hypotonic

_____14.   endocytosis

_____15.   exocytosis

a.   endocytosis of dissolved substances

b.   endocytosis of large particles

c.   movement down a concentration gradient

d.   enzyme associated with membrane pump

e.   carrier-assisted movement along a concentration gradient

f.   carrier-assisted movement against a concentration gradient

g.   solute concentration outside cell is equal to cellular solute concentration

h.   solute concentration outside cell is less than cellular solute concentration

i.   solute concentration outside cell is more than the cellular solute concentration

j.   a membrane component necessary for active transport

k.   pinocytosis and phagocytosis

l.   diffusion of water across a selectively permeable membrane

m.   this involves fusion of cytoplasmic vesicles with the plasma membrane

n.   occurs when water moves into cells having a cell wall

o.   a property of plasma membranes

Multiple Choice Questions

1. Which part of the plasma membrane makes it permeable to lipid soluble substances?
   a. phospholipids
   b. carrier proteins
   c. carbohydrates
   d. nucleic acids
   e. water

2. Which makes the plasma membrane selectively permeable to lipid insoluble substances?
   a. phospholipids
   b. carbohydrates
   c. nucleic acids
   d. carrier proteins
   e. water

3. Which DOES NOT describe simple diffusion across a plasma membrane?
   a. The membrane must be impermeable to the diffusing substance.
   b. Diffusion occurs down a concentration gradient.
   c. Diffusion of nonpolar molecules may occur through the phospholipid bilayer.
   d. Diffusion may occur through protein channels.
   e. The diffusing particle may be lipid soluble or lipid insoluble.

4. Which DOES NOT describe simple diffusion across a plasma membrane?
   a. Large particles diffuse more slowly than small particles with similar physical properties.
   b. Diffusion may be regulated by gated channels.
   c. Particles only move in one direction across the membrane.
   d. Diffusion of ions may occur through protein channels.
   e. The rate of diffusion depends on several factors.

5. Which describes osmosis in animal cells?
   a. A membrane is not necessary.
   b. It involves diffusion of solute only.
   c. The solutes in both compartments must be the same.
   d. The direction is always from the hypotonic to the hypertonic compartment.
   e. The membrane must be impermeable to all solutes.

6. Which describes osmosis in plant cells?
   a. The cell wall must be impermeable to all solutes.
   b. It involves diffusion of water down its concentration gradient.
   c. The direction of osmosis is from the solution with the higher solute concentration to the solution with the lower solute concentration.
   d. Plasmolysis occurs when turgor pressure increases.
   e. The rate of osmosis will increase as turgor pressure increases.

7. A red blood cell has a solute concentration of 0.9%. What will happen if it is placed in a 0.8% salt solution?
   a. The red blood cell will shrink if its membrane is permeable to salt and water.
   b. The red blood cell will shrink if its membrane is impermeable to salt and permeable to water.
   c. The red blood cell will swell and probably burst if its membrane is permeable to salt and water.
   d. The red blood cell will swell and probably burst if its membrane is impermeable to salt and permeable to water.
   e. The red blood cell will shrink regardless of its permeability to salt and water.

8. A red blood cell has a solute concentration of 0.9%. What will happen if it is placed in a 8.0% salt solution?
    a. The red blood cell will shrink if its membrane is permeable to salt and water.
    b. The red blood cell will shrink if its membrane is impermeable to salt and permeable to water.
    c. The red blood cell will swell and probably burst if its membrane is permeable to salt and water.
    d. The red blood cell will swell and probably burst if its membrane is impermeable to salt and permeable to water.
    e. The red blood cell will shrink regardless of its permeability to salt and water.

9. What will happen to a plant cell placed in a hypotonic sugar solution?
    a. The plant cell will swell and probably burst if its membrane is permeable to sugar and water.
    b. The plant cell will swell and probably burst if its membrane is impermeable to sugar and permeable to water.
    c. The plant cell will plasmolyze if its membrane is permeable to sugar and water.
    d. The plant cell will plasmolyze if its membrane is impermeable to sugar and permeable to water.
    e. The turgor pressure in the cell will increase.

10. What will happen to a plant cell placed in a hypertonic sugar solution?
    a. The plant cell will plasmolyze if its membrane is permeable to sugar and water.
    b. The plant cell will plasmolyze if its membrane is impermeable to sugar and permeable to water.
    d. The plant cell will swell and probably burst if its membrane is permeable to sugar and water.
    c. The plant cell will swell and probably burst if its membrane is impermeable to sugar and permeable to water.
    e. The turgor pressure in the cell will increase.

11. How are facilitated diffusion and active transport similar?
    a. Both require ATP hydrolysis to supply the energy for transport.
    b. Both are responsible for the transport of water across the membrane.
    c. Both use membrane bound carrier proteins.
    d. Both work along a concentration gradient.
    e. Both work against a concentration gradient.

12. How are facilitated diffusion and active transport different?
    a. Facilitated diffusion requires ATP to supply the energy for transport; active transport does not.
    b. Facilitated diffusion transports water; active transport does not.
    c. Facilitated diffusion works against a concentration gradient; active transport works down a concentration gradient.
    d. Facilitated diffusion works down a concentration gradient; active transport works against a concentration gradient.
    e. Facilitated diffusion uses a membrane bound carrier protein; active transport does not.

13. Which describes pinocytosis?
    a. It is a form of exocytosis.
    b. It is the ingestion of large, undissolved particles.
    c. It must work down a concentration gradient.
    d. Materials are able to cross the membrane without chemical modification.
    e. It often involves a specific membrane receptor.

14. Which describes phagocytosis?
    a. It is a form of endocytosis.
    b. It is the ingestion of dissolved materials.
    c. It is used for secretion of substances.
    d. Materials are able to cross the membrane without chemical modification.
    e. It requires a membrane bound carrier protein.

15. Some white blood cells protect us from infectious bacteria by transporting them into the white blood cell's cytoplasm. By what process does this occur?
    a. simple diffusion
    b. active transport
    c. phagocytosis
    d. facilitated diffusion
    e. exocytosis

16. Once the bacterium is taken in by the white blood cell, it is digested by hydrolytic enzymes. What organelle carries these enzymes?
    a. mitochondrion
    b. ribosome
    c. lysosome
    d. rough endoplasmic reticulum
    e. smooth endoplasmic reticulum

Concept Map Construction

Construct a concept map for each group of terms. Be sure to include appropriate connector phrases. You may add other terms as necessary and use terms in the singular or plural form.

1. selectively permeable membrane, phospholipid bilayer, carrier protein, hypotonic solution, turgor pressure
2. receptor-mediated endocytosis, membrane invagination hypothesis, active transport, ATP, concentration gradient
3. simple diffusion, lysosome, exocytosis, carbon dioxide, nonpolar molecule

Chapter 8

Processing Energy: Photosynthesis and Chemosynthesis

Section Concept Map

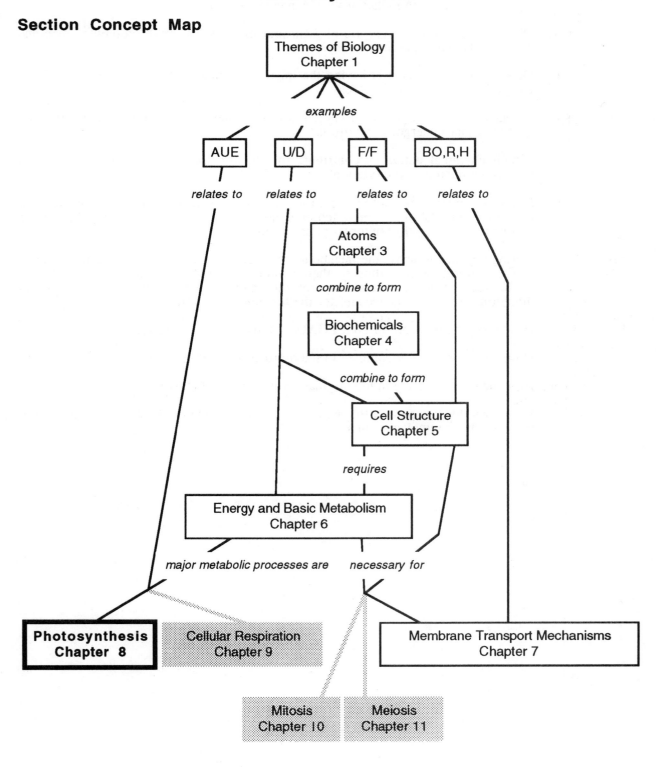

Go Figure!

Figure 8-2: ATP and NADPH are hydrolyzed and oxidized respectively in the light-independent reactions. What happens to the resulting ADP and NADP$^+$?

Figure 8-3: Why do you suppose plants use visible light for photosynthesis rather than shorter or longer wavelength radiation such as gamma rays or radio waves?

Figure 8-4: Observe the absorption spectra of plant pigments. Why are chlorophylls green and carotenoids yellow, orange or red?

Figure 8-6: Of what adaptive value is having the inner membrane of the chloroplast so highly folded?

Figure 8-8: If four ATPs are produced per O_2 released, how many ATPs are produced per water molecule that undergoes photolysis?

Which chemical activities contribute to development of the proton gradient in noncyclic photophosphorylation?

Figure 8-9: Several activities generate the proton gradient during noncyclic photophosphorylation. Which activities are *not* involved in generating the proton gradient during cyclic photophosphorylation?

Figure 8-10: During chromatography, the metabolic intermediates separate along the direction of solvent migration according to their solubilities in the solvent. A single "run" could account for the separation of the intermediates in the figure's vertical direction. How do you account for the horizontal separation?

Figure 8-11: How many carbons are in both PGALs "drained" away to produce glucose?

How does this number compare to the original input of carbons in CO_2?

How many carbons are in the original six molecules of RuBP?

How does this compare to the number of carbons in the ten PGALs remaining after two have been removed to make glucose?

Matching Questions

Write the letter of the phrase that best matches the numbered term on the left. Use each only once.

_____ 1.   chemosynthesis

_____ 2.   light-independent reactions

_____ 3.   photon

_____ 4. CO_2 fixation

_____ 5.   RuBP carboxylase

_____ 6.   PEP carboxylase

_____ 7.   autotroph

_____ 8.   heterotroph

_____ 9.   ATP synthase

_____10.   chemiosmosis

_____11.   chlorophyll

_____12.   carotenoid

_____13.   P700

_____14.   P680

_____15.   cyclic photophosphorylation

_____16.   noncyclic photophosphorylation

_____17.   photolysis

_____18.   photorespiration

_____19.   stomate

_____20.   CAM plants

a. a primary function of the light-independent reactions

b. the enzyme that catalyzes carbon fixation in C_4 plants

c. the enzyme that catalyzes ATP synthesis

d. an organism capable of photosynthesis or chemosynthesis

e. ATP and NADPH are produced

f. a drain on photosynthetic efficiency

g. autotrophic metabolism in which chemical (not light) energy is used

h. a "particle" of light

i. the enzyme that catalyzes carbon fixation in C_3 plants

j. yellow or orange plant pigment

k. an organism incapable of photosynthesis or chemosynthesis

l. carbon fixation and reduction is accomplished by these reactions

m. the process of splitting water

n. the reaction center of photosystem II

o. they fix CO_2 at night

p. a pore that allows CO_2 to enter the leaf

q. the only product is ATP

r. a green plant pigment

s. the reaction center of photosystem I

t. a model to describe the process of ATP formation

Multiple Choice Questions

1. Which DOES NOT describe autotrophs?
   a. Most get energy from the sun.
   b. Some get energy from the oxidation of chemicals.
   c. They get their organic molecules from the soil.
   d. Most release O_2 as a waste product of their activity.
   e. They are considered to be producers.

2. Which DOES NOT describe autotrophs?
   a. They comprise the majority of species.
   b. Some perform chemosynthesis.
   c. Some perform photosynthesis.
   d. They convert CO_2 into organic molecules.
   e. They are the immediate energy source for many heterotrophs.

3. Which is a reactant in photosynthesis?
   a. carbohydrate (CH_2O)
   b. water (H_2O)
   c. oxygen (O_2)
   d. methane (CH_4)
   e. ammonia (NH_3)

4. Which is a product of photosynthesis?
   a. carbon dioxide (CO_2)
   b. light
   c. water (H_2O)
   d. oxygen (O_2)
   e. methane (CH_4)

5. Which compound supplies the carbon used during photosynthesis?
   a. carbon dioxide (CO_2)
   b. light
   c. water (H_2O)
   d. oxygen (O_2)
   e. methane (CH_4)

6. Oxygen produced during photosynthesis comes from which compound?
   a. carbon dioxide (CO_2)
   b. light
   c. water (H_2O)
   d. oxygen (O_2)
   e. methane (CH_4)

7. What is accomplished in the light-dependent reactions of photosynthesis?
   a. ATP is produced.
   b. NADPH is produced.
   c. Carbohydrate is produced.
   d. ATP and NADPH are produced.
   e. ATP, NADPH and carbohydrate are produced.

8. What is accomplished in the light-independent reactions of photosynthesis?
   a. ATP is produced.
   b. NADPH is produced.
   c. Carbohydrate is produced.
   d. ATP and NADPH are produced.
   e. ATP, NADPH and carbohydrate are produced.

9. Which DOES NOT describe photons of light?
    a. They are units of light energy.
    b. Long wavelength photons contain less energy than short wavelength photons.
    c. Each pigment absorbs photons of a specific wavelength.
    d. The energy of the photon moves electrons to an orbital closer to the atomic nucleus.
    e. Only entire photons are absorbed by a pigment.

10. Which DOES NOT describe plant pigments?
    a. Chlorophylls contain magnesium.
    b. Chlorophylls absorb red and blue light.
    c. Chlorophylls are the primary plant pigment.
    d. Carotenoids absorb wavelengths not efficiently absorbed by chlorophylls.
    e. Carotenoids reflect green light.

11. Which DOES NOT describe Photosystem I?
    a. P700 is the reaction center.
    b. The antenna pigments include chlorophylls.
    c. The antenna pigments are embedded in the thylakoid membranes.
    d. There is a single Photosystem I in each chloroplast.
    e. Antenna pigments absorb light with wavelengths between 400 and 700 nm.

12. Which DOES NOT describe Photosystem II?
    a. P700 is the reaction center.
    b. The antenna pigments include carotenoids.
    c. The antenna pigments are embedded in the thylakoid membranes.
    d. There are many Photosystems II in each chloroplast.
    e. Antenna pigments absorb light with wavelengths between 400 and 700 nm.

13. Which DOES NOT describe Photosystem II?
    a. It is closely associated with oxygen production.
    b. It is responsible for increasing hydrogen ion concentration in the thylakoid lumen.
    c. It participates in photolysis.
    d. It receives electrons from water molecules.
    e. It transfers electrons directly to $NADP^+$.

14. Which DOES NOT describe Photosystem I?
    a. It receives electrons from Photosystem II via an electron transport chain.
    b. It donates the electrons used in $NADP^+$ reduction.
    c. It is responsible for decreasing hydrogen ion concentration in the stroma of the chloroplast.
    d. It contains the ATP synthase involved in ATP production.
    e. P700 receives electrons from P680.

15. What is produced during noncyclic photophosphorylation?
    a. ATP
    b. NADPH
    c. H_2O
    d. ATP and NADPH
    e. ATP, NADPH and H_2O

16. What is produced during cyclic photophosphorylation?
    a. ATP
    b. NADPH
    c. H_2O
    d. ATP and NADPH
    e. ATP, NADPH and H_2O

17. The term "chemiosmosis" refers to the production of
    a. ATP.
    b. NADPH.
    c. carbohydrate.
    d. O_2.
    e. water.

18. The term "photophosphorylation" refers to the production of
    a. ATP.
    b. NADPH.
    c. carbohydrate.
    d. O_2.
    e. water.

19. Which DOES NOT contribute to the proton gradient created during the light dependent reactions of photosynthesis?
    a. impermeability to protons of the thylakoid membrane
    b. splitting of water
    c. reduction of $NADP^+$
    d. movement of protons associated with electron transport
    e. oxidation of NADPH

20. Which DOES NOT describe the ATP synthase in chloroplasts?
    a. It is not attached to a membrane.
    b. The stalked portion forms a proton channel.
    c. The active site has ADP and P_i as substrates.
    d. It requires diffusing protons to drive synthesis.
    e. It is a structurally complex enzyme.

21. Which DOES NOT describe C_3 synthesis?
    a. The reactions are light dependent.
    b. CO_2 is attached to a five-carbon molecule.
    c. Ribulose biphosphate is involved in carbon fixation.
    d. NADPH from the light dependent reactions is a reactant.
    e. The pathway is cyclic.

22. Which DOES NOT describe C_3 synthesis?
    a. The first identifiable product of carbon fixation is a three-carbon compound.
    b. RuBP carboxylase is the enzyme of carbon fixation.
    c. ATP from the light dependent reactions is a reactant.
    d. The pathway is used by CAM and C_4 plants.
    e. Glucose is an intermediate of the Calvin-Benson Cycle.

23. Which describes C_4 plants?
    a. They are usually out-competed by C_3 plants in warm temperatures.
    b. They are adapted to a hot, dry environment.
    c. Carbon fixation occurs at night.
    d. Their carbon-fixing enzyme is less efficient at low CO_2 concentrations than the one used by C_3 plants.
    e. Carbon reduction occurs in mesophyll cells.

24. Which describes CAM plants?
    a. Carbon fixation occurs at night.
    b. The first identifiable product of carbon fixation is a three-carbon compound.
    c. They comprise a majority of plants.
    d. They are found in cold, moist environments.
    e. Their carbon-fixing enzyme is less efficient at low CO_2 concentrations than the one used by C_3 plants.

Concept Map Construction

Construct a concept map for each group of terms. Be sure to include appropriate connector phrases. You may add other terms as necessary and use terms in the singular or plural form.

1.   chemosynthesis, eukaryote, NADPH, RuBP carboxylase, mitochondrion
2.   photosynthesis, ATP, P700, water, active site
3.   ATP synthase, light dependent reactions, CAM plant, chlorophyll, diffusion

Chapter 9

Processing Energy: Fermentation and Respiration

Section Concept Map

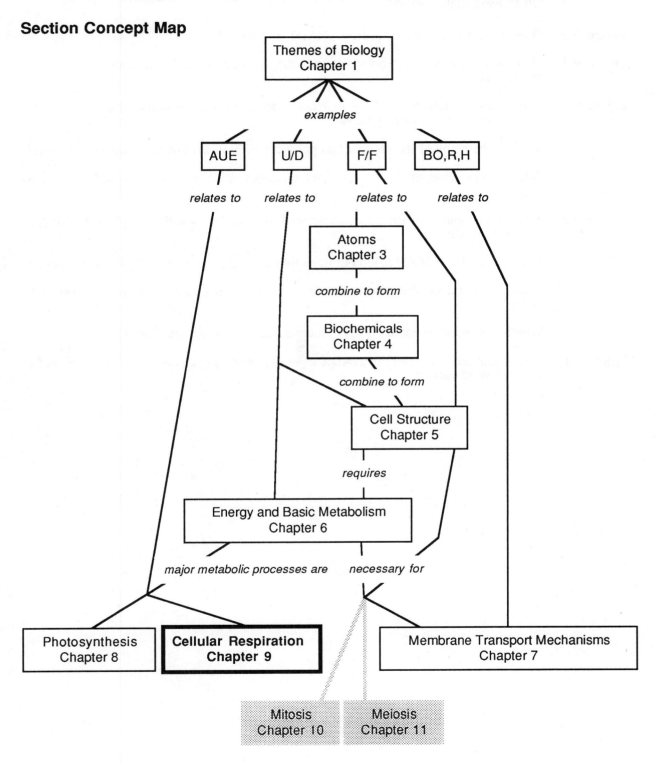

Themes of Biology
Chapter 1

examples

AUE   U/D   F/F   BO,R,H

relates to   *relates to*   *relates to*   *relates to*

Atoms
Chapter 3

combine to form

Biochemicals
Chapter 4

combine to form

Cell Structure
Chapter 5

requires

Energy and Basic Metabolism
Chapter 6

major metabolic processes are   *necessary for*

Photosynthesis
Chapter 8

**Cellular Respiration
Chapter 9**

Membrane Transport Mechanisms
Chapter 7

Mitosis
Chapter 10

Meiosis
Chapter 11

Go Figure!

Figure 9-2:   Where do the two 1,3 Diphosphoglycerates (DPG) come from?

Where do the phosphates on 1,3-Diphosphoglycerate and Phosphoenolpyruvate (PEP) eventually end up?

Figure 9-3:   Which has more potential energy: NAD^+ or NADH?

Figure 9-4:   How is it possible to produce a four-carbon butanol by fermenting three-carbon pyruvic acids?

Figure 9-5:   How does the number of CO_2 molecules produced during respiration compare to the number of carbons in glucose?

What is the name of the process that produces ATP in the electron transport system?

What is the name of the process that produces ATP in glycolysis and the Krebs cycle?

Figure 9-7:   What is the source of the Oxaloacetic acid that combines with Acetyl CoA to begin the Krebs cycle?

Succinic acid and Malic acid both have four carbons. In what way do they differ?

Malic acid and Oxaloacetic acid both have four carbons. In what ways do they differ?

Which has more potential energy: Succinic acid or Oxaloacetic acid?

Figure 9-10:   Notice that the ATP yield of NADH from glycolysis is shown as 4-6 ATP. Explain why the yield varies.

Matching Questions

Write the letter of the phrase that best matches the numbered term on the left. Use each only once.

_____ 1.   cyanobacteria

_____ 2.   anaerobe

_____ 3.   aerobe

_____ 4.   mitochondrion

_____ 5.   cytochrome oxidase

_____ 6.   ATP synthase

_____ 7.   alcoholic fermentation

_____ 8.   lactic acid fermentation

_____ 9.   substrate-level phosphorylation

_____ 10.   Krebs cycle

_h_ 11.   aerobic respiration

_____ 12.   FAD

_____ 13. NAD^+

_____ 14.   glycolysis

_____ 15.   acetyl coenzyme A

a.   organism that survives in the absence of oxygen

b.   organism that requires oxygen

c.   the products are ethyl alcohol and CO_2

d.   ATP production associated directly with a carbon compound

e.   a coenzyme used in glycolysis and Krebs cycle

f.   prokaryotes that evolved O_2 producing photosynthesis

g.   the enzymes for this are in the mitochondrial matrix

h.   oxygen is a reactant in the final step

i.   a coenzyme used only in the Krebs cycle

j.   carbon compound that enters the Krebs cycle

k.   site of aerobic respiration

l.   the conversion of glucose to two pyruvic acids

m.   enzyme responsible for ATP synthesis associated with electron transport chain

n.   pathway used by exercising muscles to oxidize NADH

o.   enzyme that reduces O_2 to water

Multiple Choice Questions

1. Which DOES NOT describe oxygen?
    a. It is necessary for the survival of anaerobes.
    b. It became abundant in the atmosphere between 2 and 3 billion years ago.
    c. It is destructive to organic chemicals.
    d. Its presence provides an opportunity to extract more energy from organic molecules.
    e. It is not required by all organisms.

2. Which DOES NOT describe aerobes?
    a. They require oxygen.
    b. They have protective enzymes that detoxify the superoxide radical.
    c. They were among the earliest forms of life to evolve.
    d. They use oxygen to release energy from food molecules.
    e. They include animals as well as plants.

3. Which DOES NOT describe glycolysis?
    a. It occurs in the cytoplasm.
    b. It begins with a phosphorylation of glucose.
    c. It produces CO_2 as a product.
    d. It is used by organisms that respire.
    e. It is used by organisms that ferment.

4. Which DOES NOT describe glycolysis?
    a. It is used to make ATP by substrate-level phosphorylation.
    b. Two three-carbon compounds are produced from each glucose.
    c. It requires O_2.
    d. It is used by prokaryotic cells.
    e. It is used by eukaryotic cells.

5. Which is a reactant in glycolysis?
    a. ADP
    b. CO_2
    c. H_2O
    d. pyruvic acid
    e. O_2

6. Which is a reactant in glycolysis?
    a. $NADP^+$
    b. NAD^+
    c. RuBP
    d. PGAL
    e. NADPH

7. Which is a (net) product of glycolysis?
    a. O_2
    b. ADP
    c. glucose
    d. $NADP^+$
    e. pyruvic acid

8. Which is a (net) product of glycolysis?
    a. CO_2
    b. NADH
    c. PGAL
    d. RuBP
    e. H_2O

9. Which describes fermentation?
    a. All fermenters produce the same end products.
    b. It requires oxygen.
    c. It produces two ATP per glucose.
    d. It is more efficient than aerobic respiration.
    e. No fermenting organisms have survived in the aerobic environment.

10. Which describes fermentation?
    a. All fermenters oxidize NADH.
    b. Most of the energy in glucose is transferred to bonds of ATP during fermentation.
    c. All ATP produced during fermentation comes from reactions involving oxidation of pyruvic acid.
    d. Ethyl alcohol is always a product of fermentation.
    e. The majority of species currently alive are only capable of fermentation.

11. Which DOES NOT describe aerobic respiration?
    a. An electron transport chain is involved.
    b. The reactions occur in the cytoplasmic fluid.
    c. It transfers more energy from glucose to the bonds of ATP than fermentation.
    d. It uses a cyclic metabolic pathway.
    e. Coenzyme A is used to make acetyl CoA.

12. Which DOES NOT describe aerobic respiration?
    a. Some ATP is produced by substrate phosphorylation.
    b. The Krebs cycle occurs in the mitochondrial matrix.
    c. Pyruvic acid from glycolysis is completely oxidized in the reactions of aerobic respiration.
    d. The enzymes for Krebs cycle are soluble.
    e. The Krebs cycle must be run twice in order to completely oxidize glucose.

13. Which is a reactant in the Krebs cycle?
    a. CO_2
    b. acetyl CoA
    c. NADPH
    d. $FADH_2$
    e. glucose

14. Which is a reactant in the Krebs cycle?
    a. O_2
    b. ATP
    c. NAD^+
    d. pyruvic acid
    e. lactic acid

15. Which is a product of the Krebs cycle?
    a. $FADH_2$
    b. ADP
    c. phosphoenol pyruvic acid
    d. acetyl CoA
    e. NAD^+

16. Which is a product of the Krebs cycle?
    a. $NADP^+$
    b. O_2
    c. pyruvic acid
    d. lactic acid
    e. CO_2

17. Which DOES NOT describe the mitochondrial electron transport system?
    a. The components are located in mitochondria.
    b. The electron carriers must be bound to membrane.
    c. It is responsible for producing the majority of ATP in aerobic respiration.
    d. It is found only in eukaryotic cells.
    e. Electron transport produces a proton gradient.

18. Which DOES NOT describe the mitochondrial electron transport system?
    a. The F_1 portion of the ATP synthase has the active site.
    b. Cytochrome oxidase is the first electron carrier of the electron transport system.
    c. ATP production relies on the process of chemiosmosis.
    d. The outer compartment of the mitochondrion has a high concentration of protons.
    e. The components of the electron transport system are in the mitochondrial matrix.

19. Which is a reactant in the mitochondrial electron transport system?
    a. pyruvic acid
    b. NADH
    c. H_2O
    d. acetyl CoA
    e. lactic acid

20. Which is a reactant in the mitochondrial electron transport system?
    a. O_2
    b. $FADH_2$
    c. CO_2
    d. O_2 and $FADH_2$
    e. O_2, $FADH_2$ and CO_2

21. Which is a product of the mitochondrial electron transport system?
    a. pyruvic acid
    b. NADH
    c. H_2O
    d. acetyl CoA
    e. lactic acid

22. Which is a product of the mitochondrial electron transport system?
    a. O_2
    b. $FADH_2$
    c. CO_2
    d. FAD
    e. $FADH_2$ and CO_2

23. How many ATP can be produced per NADH oxidized in the mitochondrial electron transport system with oxygen as the final electron acceptor?
    a. 1
    b. 2
    c. 3
    d. 24
    e. 36

24. How many ATP can be produced per $FADH_2$ oxidized in the mitochondrial electron transport system with oxygen as the final electron acceptor?
    a. 1
    b. 2
    c. 3
    d. 24
    e. 36

25.   Which can be synthesized from intermediates of glycolysis or the Krebs cycle?
    a.   carbohydrates
    b.   fatty acids
    c.   amino acids
    d.   nucleotides
    e.   All of these can be synthesized from intermediates of glycolysis or the Krebs cycle.

26.   Which can be oxidized in glycolysis or the Krebs cycle?
    a.   carbohydrates
    b.   fatty acids
    c.   amino acids
    d.   nucleotides
    e.   All of these can be oxidized in glycolysis or the Krebs cycle.

Concept Map Construction

Construct a concept map for each group of terms. Be sure to include appropriate connector phrases.
You may add other terms as necessary and use terms in the singular or plural form.

    1.   NADH, active site, proton gradient, protein, oxidation
    2.   carbohydrate, mitochondrion, photosystem I, anabolism, cilia
    3.   hypothesis, aerobic respiration, substrate-level phosphorylation, fermentation, endocytosis

Chapter 10

Cell Division: Mitosis

Section Concept Map

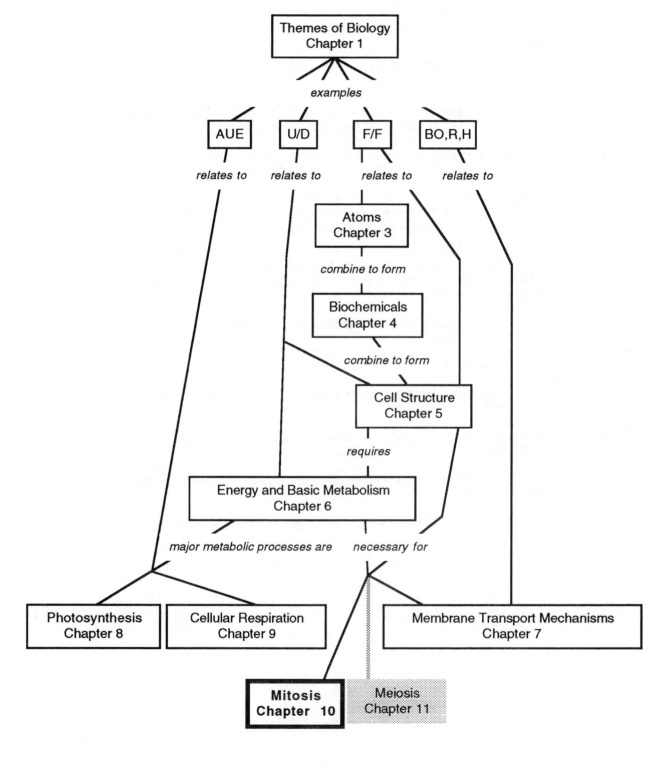

Themes of Biology
Chapter 1

examples

AUE   U/D   F/F   BO,R,H

relates to   *relates to*   *relates to*   *relates to*

Atoms
Chapter 3

combine to form

Biochemicals
Chapter 4

combine to form

Cell Structure
Chapter 5

requires

Energy and Basic Metabolism
Chapter 6

major metabolic processes are   *necessary for*

Photosynthesis
Chapter 8

Cellular Respiration
Chapter 9

Membrane Transport Mechanisms
Chapter 7

Mitosis
Chapter 10

Meiosis
Chapter 11

Go Figure!

Figure 10-2:   At which labeled structure is new DNA being made?

What is the function of the attachment site in prokaryotic fission?

Figure 10-3b: List features visible in the photographs you could use to determine which chromosomes are homologous.

Other than size, how does the homologous pair in row one, column one differ from other pairs in row one?

Other than size, how does the homologous pair in row three, column one differ from the pair in row three, column four?

All of the chromosome pairs are homologous except the X and Y pair shown in the lower right. Why are they grouped if they aren't homologous?

Figure 10-4:   Look at the nucleus third down from the on the left. What are the lines running from right to left? In what visible way are the daughter cells similar?

Compare mitosis and meiosis by looking at the four nuclei shown fourth down from the top. How are the mitotic and meiotic nuclei different? How are they similar? How are the mitotic and meiotic chromosomes different?...similar?

Now compare mitosis and meiosis by looking at the mitotic nuclei fourth down from the top and the meiotic nuclei on the bottom. How are the mitotic and meiotic nuclei different? How are they similar? How are the mitotic and meiotic chromosomes different?...similar?

Figure 10-5:   Does the size of each colored portion for a particular phase represent the length of time a "typical" cell spends in that stage?

Is the sequence from G_1 to S to G_2 or G_2 to S to G_1?

According to the diagram, in which specific phase does cytokinesis begin? In which does it end?

Figure 10-6:   In which stage do chromosomes appear thickest? They appear thicker at this stage than during the next one. What has made this change?

In which stage do chromosomes appear to be in confused disarray like a plate of spaghetti pasta?

At which stage are chromosomes "pointing" at each other like fingers from across the cell?

At which stage are there two indiscrete clumps of chromosomes?

Matching Questions

Write the letter of the phrase that best matches the numbered term on the left. Use each only once.

_____ 1. mother cell

_____ 2. prokaryotic fission

_____ 3. cytokinesis

_____ 4. chromatid

_____ 5. kinetochore

_____ 6. mitosis

_____ 7. meiosis

_____ 8. G_1

_____ 9. M phase

_____ 10. diploid

_____ 11. X and Y

_____ 12. X and X

_____ 13. asexual reproduction

_____ 14. S

_____ 15. mitotic chromosome

_____ 16. spindle apparatus

_____ 17. pericentriolar material (PCM)

_____ 18. astral spindle fibers

_____ 19. metaphase

_____ 20. cell plate

a. nonhomologous chromosome pair

b. consists of mitosis and cytokinesis

c. process produces diploid cells from diploid cells

d. contains homologous chromosome pairs

e. a dividing cell

f. a strand of densely compacted DNA and protein attached at a centromere to another one like itself

g. noncytokinetic partitioning of cellular DNA by plasma membrane growth

h. coiled, densely compacted DNA and protein, but not attached to another one like itself

i. includes polar and chromosomal fibers

j. formed by fusion of secretion vesicle membranes and their polysaccharide contents

k. radiate out in all directions from the pericentriolar material

l. phase of eukaryotic DNA synthesis

m. the physical splitting of a eukaryotic cell into two parts

n. chromosomes are aligned midway between spindle poles during this phase

o. general name for all examples of reproduction without fertilization

p. plays a key role in spindle fiber assembly

q. precedes S

r. chromosomal spindle fiber attachment site

s. process produces haploid cells from diploid cells

t. homologous chromosome pair

Multiple Choice Questions

1. Cells which have one of each kind of chromosome are called
    a. prokaryotic.
    b. haploid.
    c. diploid.
    d. pericentriolar.
    e. mitotic.

2. Cells which have two of each kind of chromosome are called
    a. prokaryotic.
    b. haploid.
    c. diploid.
    d. pericentriolar.
    e. mitotic.

3. Which statement DOES NOT describe mitosis?
    a. It is a type of division that produces identical cells.
    b. It is a type of division that may produce generation after generation of identical cells.
    c. In some multicellular organisms, it allows for asexual reproduction.
    d. It is a process which duplicates the hereditary material.
    e. In multicellular organisms, it is responsible for growth.

4. Which statement DOES NOT describe mitosis?
    a. It produces cells with a single molecule of DNA in each chromosome.
    b. It produces two cells with an equal number of chromosomes.
    c. It produces cells with chromosomes composed of two chromatids.
    d. In multicellular organisms, it is responsible for cell replacement.
    e. Chromosomal spindle fibers separate chromatids.

5. Which describes meiosis?
    a. Diploid mother cells produce diploid daughter cells.
    b. Haploid mother cells produce haploid daughter cells.
    c. Two daughter cells are produced from one mother cell.
    d. Diploid mother cells produce haploid daughter cells.
    e. Cytokinesis seldom occurs during or after meiosis.

6. Which describes meiosis?
    a. Haploid mother cells produce diploid daughter cells.
    b. Four daughter cells are produced from one mother cell.
    c. It produces cells with a chromosome number equal to the mother cell.
    d. It produces cells with one half the chromosome number of the mother cell, each chromosome composed of two chromatids.
    e. In some multicellular organisms, it allows for asexual reproduction.

7. A diploid cell with 24 chromosomes undergoes mitosis. It produces daughter cells with
    a. 12 chromosomes each containing two chromatids.
    b. 12 chromosomes each containing a single molecule of DNA.
    c. 48 chromosomes each containing a single molecule of DNA.
    d. 24 chromosomes each containing two chromatids.
    e. 24 chromosomes each containing a single molecule of DNA.

8. A haploid cell with 24 chromosomes undergoes mitosis. It produces daughter cells with
    a. 12 chromosomes each containing two chromatids.
    b. 12 chromosomes each containing a single molecule of DNA.
    c. 48 chromosomes each containing a single molecule of DNA.
    d. 24 chromosomes each containing two chromatids.
    e. 24 chromosomes each containing a single molecule of DNA.

9. A diploid cell with 24 chromosomes undergoes meiosis. It produces daughter cells with
    a. 12 chromosomes each containing two chromatids.
    b. 12 chromosomes each containing a single molecule of DNA.
    c. 48 chromosomes each containing a single molecule of DNA.
    d. 24 chromosomes each containing two chromatids.
    e. 24 chromosomes each containing a single molecule of DNA.

10. A diploid cell with 36 chromosomes undergoes meiosis. It produces daughter cells with:
    a. 18 chromosomes each containing two chromatids.
    b. 18 chromosomes each containing a single molecule of DNA.
    c. 72 chromosomes each containing a single molecule of DNA.
    d. 36 chromosomes each containing two chromatids.
    e. 36 chromosomes each containing a single molecule of DNA.

11. All the cells of a multicellular plant have been observed to contain 27 chromosomes. This plant is composed of
    a. prokaryotic, diploid cells.
    b. prokaryotic, haploid cells.
    c. eukaryotic, haploid cells.
    d. eukaryotic, diploid cells.
    e. meiotic daughter cells prior to undergoing mitosis.

12. A single-celled organism has been observed to contain a single circular strand of DNA. This organism is
    a. a prokaryotic, diploid cell.
    b. a prokaryotic, haploid cell.
    c. a eukaryotic, haploid cell.
    d. a eukaryotic, diploid cell.
    e. a meiotic daughter cell.

13. Most persons have a total of 46 chromosomes in each body cell. Which is true of these chromosomes?
    a. Twenty-three came from the father and 23 from the mother.
    b. Twenty-two came from the father and 24 from the mother.
    c. All 46 came from the mother.
    d. All 46 came from the father.
    e. The number contributed by each parent is highly variable.

14. Most persons have a total of 46 chromosomes in each body cell. Which is true of these chromosomes?
    a. This is the human haploid number of chromosomes.
    b. This is the human diploid number of chromosomes.
    c. Fertilization will restore the number to 92.
    d. Twenty-three came from the paternal grandfather and 23 from the maternal grandfather.
    e. Twenty-two came from the paternal grandfather and 24 from the maternal grandmother.

15.   In which phase does chromosome condensation occur?
       a. interphase
       b. prophase
       c. metaphase
       d. anaphase
       e. telophase

16.   In which phase does chromosomal spindle fiber attachment occur?
       a. interphase
       b. prophase
       c. metaphase
       d. anaphase
       e. telophase

17.   Which does the statement describe? These spindle fibers run from the pericentriolar material on one
       side of a cell to the pericentriolar material at the other.
       a. astral spindle fibers
       b. polar spindle fibers
       c. chromosomal spindle fibers
       d. spindle fibers in animal cells
       e. spindle fibers in plant cells

18.   Which does the statement describe? These spindle fibers run from pericentriolar material to a
       kinetochore.
       a. astral spindle fibers
       b. polar spindle fibers
       c. chromosomal spindle fibers
       d. spindle fibers in animal cells
       e. spindle fibers in plant cells

19.   What event marks the beginning of anaphase?
       a. condensation of chromosomes
       b. alignment of chromosomes at the metaphase plate
       c. synchronous separation of sister chromatids
       d. chromosomes reach their respective spindle poles
       e. chromosomes uncoil

20.   What event marks the end of anaphase?
       a. the condensation of chromosomes
       b. the alignment of chromosomes at the metaphase plate
       c. the synchronous separation of sister chromatids
       d. chromosomes reach their respective spindle poles
       e. chromosomes uncoil

21.   Cytokinesis in animals results from
       a. contraction of a ring of microtubules.
       b. construction of a new cell membrane and cell wall.
       c. contraction of a ring of microfilaments
       d. construction of a new nuclear membrane.
       e. construction of a cell wall around each new nucleus.

22.   Cytokinesis in plants results from
       a. contraction of a ring of microtubules.
       b. construction of a new cell membrane and cell wall.
       c. contraction of a ring of microfilaments
       d. construction of a new nuclear membrane.
       e. construction of a cell wall around each new nucleus.

Concept Map Construction

Construct a concept map for each group of terms. Be sure to include appropriate connector phrases. You may add other terms as necessary and use terms in the plural or singular form.

1. mitosis, DNA, daughter cell, nucleus, prokaryotic fission
2. microtubule, chromosome, chromatid, anaphase, pericentriolar material
3. sexual reproduction, meiosis, haploid, cytokinesis, homologous chromosome

Chapter 11

Cell Division: Meiosis

Section Concept Map

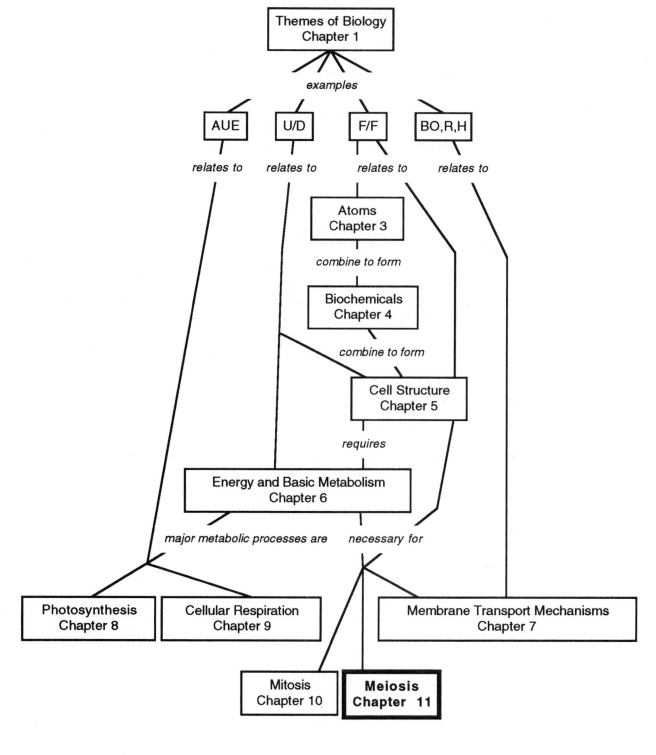

Go Figure!

Figure 11-1:   What process reduces the cellular chromosome number to half the original number in animals?...in plants?

What process doubles the cellular chromosome number in animals?...in plants?

Is a spore haploid or diploid? Name plant haploid structures in which mitosis occurs?

What do repeated mitotic divisions of an animal zygote produce?...of a plant zygote?

Name one type of cell produced by mitosis in a pollen grain.

Figure 11-2:   One pair of homologous chromosomes is shown in blue. What is the significance of one member of the pair being shown in dark blue and the other in light blue?

What is separated during Anaphase I?...during Anaphase II?

Is DNA duplicated during the blue-shaded interphase? How can you tell?

Figure 11-3:   The shading used in this diagram differs from that used in Figure 11-2. What is the significance of shading one chromosome of the homologous pair blue and the other orange?

Assume these are chromosomes in the cells of a new multicellular animal. If the blue chromosome came from the father, where did the orange chromosome from?

Figure 11-4:   It usually is not possible to look at chromosomes under the microscope and tell which gene forms are physically located on a chromosome. Since this is the case, how would you be able to tell if a plant chromosome had the gene for tall <u>and</u> the gene for white flowers on the same chromosome?

Figure 11-6:   The shading of chromosomes in this figure is similar to Figure 11-3. What is the significance of shading one chromosome of a homologous pair blue and the other orange?

What is the significance of having nonhomologous chromosomes shaded the same color?

Is the cell represented an animal or a plant cell? How can you tell?

What determines which arrangement of blue and orange chromosomes a cell will have? Can you predict the outcome with certainty?

Matching Questions

Write the letter of the phrase that best matches the numbered term on the left. Use each only once.

_____ 1. zygote

_____ 2. life cycle

_____ 3. synaptonemal complex

_____ 4. spore

_____ 5. germ cell

_____ 6. somatic cell

_____ 7. genetic variability

_____ 8. reduction division

_____ 9. synapsis

_____10. tetrad

_____11. crossing over

_____12. chiasma

_____13. asexual reproduction

_____14. maternal chromosome

_____15. paternal chromosome

_____16. prophase I

_____ 17. prophase II

_____18. metaphase I

_____19. metaphase II

_____20. haploid

a. may act to hold synapsing chromosomes together

b. cell of a multicellular organism capable of undergoing meiosis

c. reduces the number of chromosomes by half

d. variety in characteristics among individuals of a population

e. all events in the life of an organism from its reproductive formation until it reproduces

f. is usually diploid and formed at fertilization

g. reproduction without fertilization

h. body cell of a multicellular organism normally not capable of undergoing meiosis

i. will produce pollen by mitosis

j. an exchange of parts between homologous chromosomes

k. the member of a homologous pair contributed by a sperm

l. meiotic phase when chiasmata are visible

m. a point on a chromosome where crossing over has occurred

n. close alignment of homologs along their entire length

o. a member of a homologous pair contributed by the egg

p. random positioning of homologues occurs

q. a homologous pair of preanaphase I chromosomes

(choices continued on the next page)

r. generally applies to a cell not in the process of division which contains one of each type of chromosome present

s. immediately precedes anaphase II

t. condensing chromosomes are not paired in this phase

Multiple Choice Questions

1. The first diploid cell of a multicellular plant's life cycle is called
    a. a spore.
    b. an embryo sac.
    c. an egg.
    d. a zygote.
    e. a pollen grain.

2. The first haploid cell of a multicellular plant's life cycle is called
    a. a spore.
    b. an embryo sac.
    c. an egg.
    d. a zygote.
    e. a pollen grain.

3. What process produces the first diploid cell of a multicellular plant's life cycle?
    a. synapsis
    b. meiosis
    c. fertilization
    d. crossing over
    e. mitosis

4. What process produces the first haploid cell of a multicellular plant's life cycle?
    a. synapsis
    b. meiosis
    c. fertilization
    d. crossing over
    e. mitosis

5. Meiosis increases genetic variability. Some of this variability is produced in a zygote by
    a. mixing paternal and maternal chromosomes during fertilization.
    b. reducing the number of chromosomes after fertilization.
    c. increasing the number of chromosomes after fertilization.
    d. normal mitotic divisions.
    e. the contribution of a haploid number of chromosomes from the sperm and a diploid number by the egg.

6. Meiosis increases genetic variability. Some of this variability results from
    a. mixing paternal and maternal chromosomes during gamete formation.
    b. reducing the number of homologous pairs during meiosis.
    c. increasing the number of homologous pairs during meiosis.
    d. crossing over.
    e. splitting kinetochores during anaphase I.

7. During prophase I of meiosis,
    a. the chromosomes unwind to reform chromatin material.
    b. the chromosomes align on the metaphase plate.
    c. crossing over occurs.
    d. kinetochores split.
    e. homologous chromosomes separate.

8. During prophase I of meiosis,
    a. tetrads separate.
    b. chromosomes condense.
    c. kinetochores form.
    d. cytokinesis occurs.
    e. homologous chromosomes align on the metaphase plate.

Use the following information for questions 9 and 10.

Tomato plants have a gene for fruit shape and another gene for shape of the flower shoot, both on the same chromosome. The fruit shape gene comes in two forms: one produces round fruit, the other long fruit. The gene for the shape of the flower shoot also comes in two forms: branched and unbranched. For problems 9 and 10, assume that in a diploid plant, one homologue has the allele for round fruit and the allele for unbranched flower shoots. The homologous chromosome has the allele for long fruit and the allele for branched flower shoots.

9. If crossing over DOES NOT occur in tomatoes, what gene combination will be seen in the gametes?
    a. The round fruit allele with the branched flower shoot allele is in all gametes.
    b. Some gametes contain the round fruit allele with the branched flower shoot allele, and other gametes contain the long fruit allele with the branched flower shoot allele.
    c. Some gametes contain the round fruit allele with the unbranched flower shoot allele, and others contain the long fruit allele with the branched flower shoot allele.
    d. Some gametes contain the round fruit allele with the unbranched flower shoot allele, and others contain the long fruit allele with the unbranched flower shoot allele.
    e. Some contain the round fruit allele with the branched flower shoot allele. and others contain the long fruit allele with the unbranched flower shoot allele.

10. If crossing over DOES occur in tomatoes, what NEW gene combinations may be seen in the gametes?
    a. The round fruit allele with the branched flower shoot allele is in all gametes.
    b. Some gametes contain the round fruit allele with the branched flower shoot allele, and other gametes contain the long fruit allele with the branched flower shoot allele.
    c. Some gametes contain the round fruit allele with the unbranched flower shoot allele, and others contain the long fruit allele with the branched flower shoot allele.
    d. Some gametes contain the round fruit allele with the unbranched flower shoot allele, and others contain the long fruit allele with the unbranched flower shoot allele.
    e. Some contain the round fruit allele with the branched flower shoot allele. and others contain the long fruit allele with the unbranched flower shoot allele.

11. Meiotic chromosomes do not actually press against one another during synapsis. Instead, they are held near each other by the
    a. synaptonemal complex.
    b. kinetochore.
    c. recombination nodules.
    d. centromere.
    e. single strands of DNA extending between them.

12. The enzymes required for crossing over are found within the
    a. synaptonemal complex.
    b. kinetochore.
    c. recombination nodules.
    d. centromere.
    e. single strands of DNA extending between them.

13. During Meiosis I, the nucleolus is dispersed at
    a. prophase.
    b. metaphase.
    c. anaphase.
    d. telophase.
    e. the onset of cytokinesis.

14. During Meiosis I, chromosomal spindle fibers shorten at
    a. prophase.
    b. metaphase.
    c. anaphase.
    d. telophase.
    e. the onset of cytokinesis.

15. During anaphase II
    a. the chromosomes unwind to reform chromatin material.
    b. the chromosomes align on the metaphase plate.
    c. the nucleolus is dispersed.
    d. kinetochores split.
    e. homologous chromosomes separate.

16. During anaphase II
    a. the chromosomes begin to condense.
    b. synapsis occurs.
    c. chromosomal spindle fibers shorten.
    d. pairs of chromosomes separate.
    e. cytokinesis occurs.

17. In what way do mitosis and meiosis <u>differ</u>?
    a. Mitosis produces cells which are identical; meiosis produces cells which are not identical.
    b. Replication of DNA precedes mitosis.; it does not precede meiosis.
    c. A spindle apparatus forms during meiosis; one does not form during mitosis.
    d. Centrioles form during mitosis; they do not form during meiosis.
    e. Chromosomal spindles shorten during mitosis; they do not during meiosis.

18. In what way do mitosis and meiosis <u>differ</u>?
    a. Mitosis results in four new cells; meiosis results in two new cells.
    b. Mitosis produces diploid cells from haploid ones; meiosis produces haploid cells from diploid ones.
    c. Cytokinesis generally occurs once during or after mitosis; it usually occurs at two different points during or after meiosis.
    d. Chromosomal spindles do not shorten during mitosis; they shorten during meiosis.
    e. Kinetochores split during mitosis; they do not during meiosis.

19. In what way are mitosis and meiosis similar?
    a. Both produce two new cells from one mother cell.
    b. Both produce haploid cells from diploid cells.
    c. Chromosomal spindles shorten and polar spindles lengthen during both processes.
    d. Homologous chromosomes separate during both processes.
    e. Both result in the division of the cytoplasm into two parts.

20. In what way are mitosis and meiosis similar?
    a. Both produce four new cells from one mother cell.
    b. Both produce diploid cells from diploid cells.
    c. Chromosomal spindles lengthen and polar spindles shorten during both processes.
    d. Both processes generally take between a few minutes and one hour to complete.
    e. Kinetochores split during both processes.

Concept Map Construction

Construct a concept map for each group of terms. Be sure to include appropriate connector phrases. You may add other terms as necessary and use terms in the plural or singular form.

1. mitosis, meiosis, pollen, flowering plant, homologue
2. germ cell, zygote, crossing over, metaphase I, cytokinesis
3. anaphase I, chromosomal spindle fiber, pericentriolar material, genetic recombination, independent assortment

Chapter 12

On The Trail of Heredity

Section Concept Map

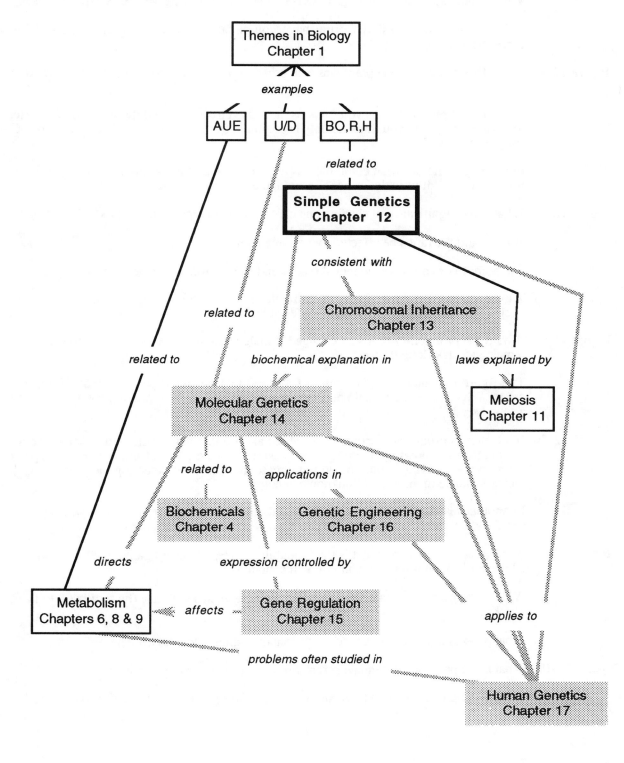

Go Figure!

Figure 12-2: What is the significance of yellow and green peas in the same pod?

What is the purpose of removing the anthers from plants before they have produced pollen?

Why would "accidental" matings have been undesirable during Mendel's experiments?

Why would "fatherless" offspring have complicated Mendel's interpretation of his experiments?

Figure 12-3: In the P_1 generation, one plant was true breeding for round seeds and one for wrinkled seeds.

Describe the physical steps you think Mendel used to cross breed these two plants (refer to Figure 12-2). Do you think he only used pollen from the round seed producing plant? Why or why not?

In the F_2 generation, what combinations of peas do you think Mendel saw in a pod?...all round seeds?...all wrinkled seeds?...wrinkled and round mixed?

Figure 12-4: What is the significance of shading one chromosome of a pair blue and the other orange?

How would the ratios be affected by crossing over?

Are both Aa offspring sets identical (*i.e.*, A and a each from the same parent)?

The new diploid combinations in the middle are found in what stage of an animal's life cycle (See Chapter 12)?...in a plant's life cycle?

Figure 12-5: Would the probability ratios be changed if male gametes were listed along the side and female gametes along the top?

The figure indicates there is a ratio of 3 round to 1 wrinkled peas produced from this cross. What is the relationship between phenotype ratio and the probability of producing a round seed?

Figure 12-7: If a testcross produces 10 peas, all of which are round, what would you conclude about the plant of unknown genotype? Is it possible that the plant of unknown genotype is heterozygous? What could you do to increase the chance of detecting the presence of a recessive allele in the plant of unknown genotype?

In another testcross, 150 peas are produced, all of which are round. Can you say with certainty that the plant of unknown genotype is homozygous dominant?

Figure 12-8: According to the figure, what is the probability an F_1 produced gamete will contain the alleles RY? What is the probability it will contain ry?

What is the probability that RY and ry containing gametes will fuse (be fertilized) to form a $RrYy$ pea?

What process insures that gametes each contain a single "R" and a single "Y?"

Figure 12-10: What is the probability the male parent will produce a gamete with the genotype ABC?

What is the probability that both parents will produce gametes with the genotype ABC?

What is the probability these parents will produce an *AaBbCc* child?

What is the probability these parents will produce a child intermediate (a 3 on the bottom scale) in skin pigmentation?

Matching Questions

Write the letter of the phrase that best matches the numbered term on the left. Use each only once.

_____ 1.   heredity

_____ 2.   blending inheritance

_____ 3.   cross

_____ 4.   testcross

_____ 5.   breed true

_____ 6.   hybrid

_____ 7.   dominant characteristic

_____ 8.   recessive characteristic

_____ 9.   "units of inheritance"

_____10.   homozygous

_____11.   phenotype

_____12.   Law of Segregation

_____13.   probability

_____14.   dihybrid cross

_____15.   incomplete dominance

_____16.   codominance

_____ 17.   epistasis

_____18.   pleiotropy

_____19.   mutation

_____20.   wild type

a.   to mate two individuals (one may or may not be a recessive phenotype))

b.   describes organisms whose offspring always have the same characteristic as the parents

c.   the parental characteristic <u>not</u> shown by a hybrid

d.   possessing two of the same alleles for a gene

e.   predicts alleles of the same gene will separate

f.   the passage of traits from one generation to the next

g.   the chance of an event over the sum of the chances for and against it

h.   mating of an organism of unknown genotype to one homozygous recessive for the same characteristic

i.   hereditary fluids mixing to form offspring

j.   offspring from parents that differ in a characteristic

k.   Mendel's name for genes

l.   the parental characteristic shown by a hybrid

m.   mating parents that differ in two characteristics

n.   the physical appearance

o.   both alleles fully expressed in the hybrid

p.   the original allele form in a population

q.   a permanent change in a gene

r.   one gene affecting many characteristics

s.   one gene masking the effect of another

t.   produces a hybrid phenotypically intermediate to the parents

Multiple Choice Questions

1. A gene exists in two different forms: *A* and *a*. An organism with the genotype *A*a will produce equal numbers of *A*-containing and *a*-containing gametes. This demonstrates
   a. the Law of Segregation.
   b. the Law of Independent Assortment.
   c. recessiveness.
   d. dominance.
   e. homozygosity.

2. Gene "A" exists in two forms: *A* and *a*. Gene "B", located on a different chromosome, also exists in two forms: *B* and *b*. An organism with the genotype *AaBb* will produce equal numbers of *AB-*, *Ab-*, *aB-*, and *ab*-containing gametes. The fate of the "A" and "B" genes relative to each other demonstrates
   a. the Law of Segregation.
   b. the Law of Independent Assortment.
   c. recessiveness.
   d. dominance.
   e. homozygosity.

3. Gene "A" exists in two forms: *A* and *a*. What are *A* and *a* called?
   a. P
   b. F_1
   c. genotypes
   d. alleles
   e. hybrids

4. Gene "A" exists in two forms: *A* and *a*. An individual's genotype is known to be *Aa*. Which applies to this individual?
   a. It is homozygous.
   b. It is heterozygous.
   c. It breeds true.
   d. It has two copies of the same allele.
   e. It has the recessive phenotype.

5. Two pea plants breed true for round peas. If they are crossed to produce large numbers of peas, you would expect to find
   a. round peas in all pods.
   b. round peas in some pods, but there is a chance others will have wrinkled peas.
   c. round peas in one half the pods, and wrinkled in the others.
   d. wrinkled peas in some pods, but there is a chance others will have round peas.
   e. wrinkled peas in some pods, round peas in some pods, and round and wrinkled peas mixed in the others.

6. Two pea plants are hybrids for pea shape. If they are crossed to produce a large number of peas, you would expect to find
   a. round peas in all pods.
   b. round peas in some pods, but there is a chance others will have wrinkled peas.
   c. round peas in one half the pods, and wrinkled in the others.
   d. wrinkled peas in some pods, but there is a chance others will have round peas.
   e. wrinkled peas in some pods, round peas in some pods, and round and wrinkled peas mixed in the others.

7. Two pea plants breed true: one for round peas, the other for wrinkled. If pollen from the round pea plant is transferred to the wrinkled pea plant, what would you expect the peas to look like?
   a. round peas in all pods.
   b. round peas in some pods, but there is a chance others will have wrinkled peas.
   c. round peas in one half the pods, and wrinkled in the others.
   d. wrinkled peas in some pods, but there is a chance others will have round peas.
   e. wrinkled peas in all pods.

8. Two pea plants breed true: one for round peas, the other for wrinkled. If pollen from the wrinkled pea plant is transferred to the round pea plant, what would you expect the peas to look like?
   a. round peas in all pods.
   b. round peas in some pods, but there is a chance others will have wrinkled peas.
   c. round peas in one half the pods, and wrinkled in the others.
   d. wrinkled peas in some pods, but there is a chance others will have round peas.
   e. wrinkled peas in all pods.

9. Two pea plants breed true: one for tall, the other for dwarf. The plants are crossed, and all hybrid offspring are tall. Which applies to the tall allele?
   a. It is codominant.
   b. It is dominant.
   c. It is incompletely dominant.
   d. It is recessive.
   e. It is epistatic.

10. Two pea plants breed true: one for tell, the other for dwarf. The plants are crossed, and all hybrid offspring are tall. Which applies to the allele for dwarf?
    a. It is codominant.
    b. It is dominant.
    c. It is incompletely dominant.
    d. It is recessive.
    e. It is epistatic.

NOTE: Answers for questions 11 through 14 are provided in the Appendix. Questions 15 through 18 constitute a similar sequence, but the answers are not given.

Use the following information for questions 11-14.

There is a gene in summer squash plants that determines fruit color. The dominant allele (*Y*) produces yellow fruit and the recessive (*y*) results in green fruit.

11. A *YY* plant will produce _____ fruit.
    a. yellow
    b. green
    c. yellow and green
    d. white
    e. white, yellow, and green

12. What is the genotype of a plant that produces green fruit?
    a. *YY*
    b. *Yy*
    c. *yy*
    d. *YY* or *yy*
    e. *Yy* or *yy*

13. If you want to determine the genotype of a plant that produced 5 yellow fruit, with which genotype would you cross it?
    a. *YY*
    b. *Yy*
    c. *yy*
    d. *YY* or *yy*
    e. *Yy* or *yy*

14. You performed a test cross on a yellow fruit producing plant and 10 offspring all grew into plants that produced yellow fruit. Which would be the most certain conclusion?
    a. The genotype of the yellow fruit plant is *YY*.
    b. The genotype of the yellow fruit plant is *Yy*.
    c. The genotype of the yellow fruit plant is *yy*.
    d. The genotype of the yellow fruit plant is either *YY* or *Yy*.
    e. The genotype of the yellow fruit plant is either *YY* or *yy*.

Use the following information for questions 15 - 18.

The gene for pea plant height may be represented with a "T" and the gene for flower color with a "C". Then *T* represents the allele for tall and *t* the allele for dwarf; *C* represents the allele for purple flowers and *c* the allele for white flowers.

15. A *TTcc* plant will be
    a. tall with white flowers.
    b. dwarf with white flowers.
    c. tall with purple flowers.
    d. short with purple flowers.
    e. tall with purple and white flowers.

16. What is the genotype of a dwarf, white-flowered plant?
    a. *TTCc*
    b. *TtCc*
    c. *TTcc*
    d. *Ttcc*
    e. *ttcc*

17. If you want to determine the genotype of a tall, purple-flowered plant, with which genotype would you cross it?
    a. *TTCc*
    b. *TtCc*
    c. *TTcc*
    d. *Ttcc*
    e. *ttcc*

18. You performed a test cross on a tall, purple-flowered plant and 10 offspring all grew into tall, purple-flowered plants. Which would be the most certain conclusion?
    a. The genotype of the yellow fruit plant is *TTCC*.
    b. The genotype of the yellow fruit plant is *TTCc*.
    c. The genotype of the yellow fruit plant is *TtCc*.
    d. The genotype of the yellow fruit plant is either *TTCC* or *TTCc*.
    e. The genotype of the yellow fruit plant is either *TTCC*, *TTCc*, or *TtCc*.

Use the following information for questions 19 - 22.

The gene for pea plant height may be represented with a "T" and the gene for flower color with a "C". Then *T* represents the allele for tall and *t* the allele for dwarf; *C* represents the allele for purple flowers and *c* the allele for white flowers.

19.   What kind of gametes can a *TTcc* individual produce?
       a.   *TT*
       b.   *cc*
       c.   *Tc*
       d.   *Tc*, *TT*, and *cc*
       e.   *Tc*, *TC*, *TT*, and *cc*

20.   What kind of gametes can a *TtCc* individual produce?
       a.   *TT*
       b.   *Cc*
       c.   *Tc* and *tC*
       d.   *Tc*, *TT*, and *cc*
       e.   *TC*, *Tc*, *tC*, and *tc*

21.   If a cross is made between a plant of genotype *TtCc* and one of type *TTcc*, what is the probability that an offspring will be genotype *TTcc*?
       a.   3/4
       b.   1/2
       c.   1/4
       d.   1/8
       e.   1/16

22.   If a cross is made between two plants of genotype *TtCc*, what is the probability that an offspring will be genotype *TTcc*?
       b.   1/2
       c.   1/4
       d.   1/8
       e.   1/16

Use the following information for questions 23 - 26.

There is a gene in summer squash plants that determines fruit color. The dominant allele (*Y*) results in yellow fruit and the recessive (*y*) results in green fruit. A plant with the genotype *Yy* has yellowish green fruit. There is a second gene in which the production of white fruit (*W*) is dominant to the production of non-white fruit (*w*). A plant with the genotype *Ww* produces white fruit.

23.   If a *Yy* plant is self-pollinated, the results are 1/4 of the fruit are yellow, 1/2 are yellow-green, and 1/4 are green. What is the name applied to this type of allele relationship?
       a.   incomplete dominance
       b.   epistasis
       c.   recessiveness
       d.   dominance
       e.   codominance

24.   A plant with the genotype *WwYY* produces white fruit, and a *wwYY* plant produces yellow fruit. What is the name applied to this type of gene relationship?
       a.   incomplete dominance
       b.   epistasis
       c.   recessiveness
       d.   dominance
       e.   codominance

25.   A plant with the genotype *WWyy* produces white fruit, and a *wwyy* plant produces green fruit. What proportion of the offspring from a cross of a *WwYy* plant with a *wwyy* plant would produce green fruit?
a.   1 (100% or absolute certainty)
b.   1/2
c.   1/4
d.   1/8
e.   0 (Barring a mutation, it can not happen.)

26.   An individual with the genotype *WwYy* produces white fruit. An individual with the genotype *wwYy* produces yellowish green fruit. If a *WwYY* plant is crossed with a *Wwyy* plant, what proportion of their offspring is expected to produce yellow fruit?
a.   1 (100% or absolute certainty)
b.   3/4
c.   1/2
d.   1/4
e.   0 (Barring a mutation, it can not happen.)

Concept Map Construction

Construct a concept map for each group of terms. Be sure to include appropriate connector phrases. You may add other terms as necessary.

1.   hybrid, F_1, Law of Segregation, dominant, phenotype
2.   gene, allele, homologous chromosome, epistasis, codominant
3.   anaphase I, Law of Independent Assortment, discontinuous variation, polygenic inheritance, heterozygous

Chapter 13

Genes and Chromosomes

Section Concept Map

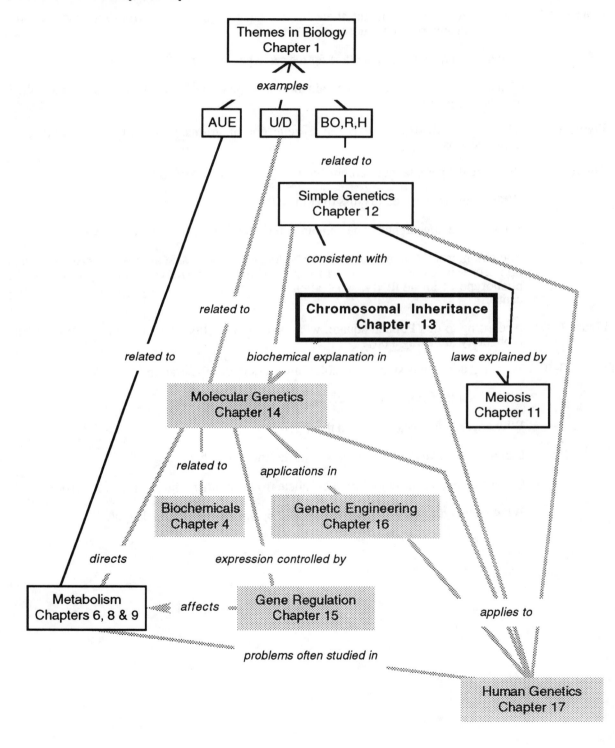

Go Figure!

Figure 13-1: Observe the larger homologous pair. What is the significance of one allele being shaded red and the other green? Why did the artist draw this pair the same size and shape?

Turn your attention to the smaller pair. What is the significance of both alleles being shaded the same color?

Figure 13-3: Can a wild type and mutant allele for the same gene occur on the same chromosome (barring an abnormal chromosome)?

Can a wild type and mutant allele for the same gene occur in the same fly?

Can a single chromosome #2 contain the wild type allele for one gene and a mutant allele for another gene?

Figure 13-4: This figure illustrates all parental gene combinations. List all nonparental gene combinations.

Figure 13-5: What are the parental gene combinations shown at the top of "b?"

Which gene is involved in cross over in "b?"

What are the nonparental gene combinations which result from crossing over?

The chromosomes in the circles at the bottom of "b" represent all possible cross over and non-cross over combinations. In order for the genes shown to be expressed in the phenotype of an adult fly, what alleles of each gene must the second chromosome contain?

Figure 13-7: According to this Punnett square, what is the probability of having a girl? What is the probability of having a boy?

Figure 13-10: Which generation is skipped by color blindness in this illustration?

What is another word for carrier?

Why are only females carriers in this figure?

Can a son receive the color blindness allele from his father?

Can a son receive the color blindness allele from a mother with normal color vision?

What phenotypes may the parents have if they have a color blind daughter?

Matching Questions

Write the letter of the phrase that best matches the numbered term on the left. Use each only once.

_____ 1. linkage group

_____ 2. mutant

_____ 3. *Drosophila* with long aristae

_____ 4. crossing over

_____ 5. map unit

_____ 6. genetic marker

_____ 7. chromosome mapping

_____ 8. mutagenic

_____ 9. polytene chromosome

_____ 10. sex chromosomes

_____ 11. autosomes

_____ 12. X chromosome

_____ 13. Y chromosome

_____ 14. muscular dystrophy

_____ 15. carrier

_____ 16. pedigree

_____ 7. Y-linked

_____ 18. gene dosage

_____ 19. chromosomal aberrations

_____ 20. oncogene

a. equal to a crossing over frequency of one percent

b. a wild type

c. using cross over frequency to determine the relative distance and position of genes

d. nonhomologous pair in human male

e. unpaired, gender determining in human males

f. describes an organism containing an inheritable change in DNA in the homozygous condition

g. unpaired, non-gender determining in human males

h. genes on the same chromosome

i. this contains numerous duplicated DNA strands lying side by side

j. X-linked characteristic

k. a family tree containing information about specific inherited traits

l. gene found exclusively on the human gender determining chromosome

m. an allele producing a visible change in phenotype from wild type

n. describes any agent which causes inheritable changes in DNA

o. cancer causing gene

p. number of functional copies of a gene

q. explains incomplete linkage

r. result only from abnormal chromosomal breakage

s. homologous pairs in human male

t. heterozygous for a recessive trait

Multiple Choice Questions

1. Groups of genes on the same chromosome form
    a. homologous pairs.
    b. sex chromosomes
    c. linkage groups.
    d. carriers.
    e. translocations.

2. During meiosis, groups of genes on the same chromosome demonstrate
    a. the Law of Segregation.
    b. the Law of Independent Assortment.
    c. dominance.
    d. linkage.
    e. heterozygosity.

Use the following information for questions 3 and 4.

The diploid number of chromosomes in the *Aedes* genus of mosquitoes is 6. There are no microscopically or genetically detectable sex chromosomes.

3. How many linkage groups are there in an *Aedes* mosquito?
    a. 12
    b. 6
    c. 3
    d. 2
    e. There are too few chromosomes in *Aedes* to have any linkage groups.

4. How many linkage groups are there in the gametes of an *Aedes* mosquito?
    a. 12
    b. 6
    c. 3
    d. 2
    e. There are too few chromosomes in *Aedes* to have any linkage groups.

Use the following information for questions 5 and 6.

Male honey bees (*Apis mellifera*) are haploid with 16 chromosomes, while the females are diploid with 32. There are no microscopically or genetically detectable sex chromosomes.

5. How many linkage groups are there in a male honey bee?
    a. 8
    b. 16
    c. 24
    d. 32
    e. 40

6. How many linkage groups are there in a female honey bee?
    a. 8
    b. 16
    c. 24
    d. 32
    e. 40

7. An exchange of genetic material between homologous chromosomes is called
   a. linkage.
   b. crossing over.
   c. translocation.
   d. duplication.
   e. inversion.

8. An exchange of genetic material between nonhomologous chromosomes is called
   a. linkage.
   b. crossing over.
   c. translocation.
   d. duplication.
   e. inversion.

NOTE: Answers for questions 9 through 11 are provided in the Appendix. Questions 12 through 14 constitute a similar sequence, but the answers are not given.

Use the following information for questions 9-11.

Genes "A" and "B" are found on the same chromosome. In one individual, allele A and b are on one homologue and a and B on the other.

9. What gametes can this organism produce in terms of "A" and "B" alleles if crossing over does not occur?
   a. *AA* and *BB*
   b. *AB* and *ab*
   c. *Ab,* and *aB*
   d. *AB, ab, Ab,* and *aB*
   e. *aa* and *BB*

10. What gametes can this organism produce in terms of "A" and "B" alleles if crossing over occurs?
    a. *AA* and *BB*
    b. *AB* and *ab*
    c. *Ab,* and *aB*
    d. *AB, ab, Ab,* and *aB*
    e. *aa* and *BB*

11. Assume crossing over does not occur. What genotypes may be seen in the offspring from a *AAbb* crossed with an *AaBb*?
    a. *AABB*
    b. *AaBb*
    c. *AABB* and *AaBb*
    d. *AABB, AABb, AAbb, AaBB, AaBb,* and *aabb*
    e. *AABB, AABb, AAbb, AaBB, AaBb, Aabb, aaBB, aaBb,* and *aabb*

Use the following information for questions 12-14.

Genes "A" and "B" are found on the same chromosome. In one individual, allele A and B are on one homologue and a and b on the other.

12. What gametes can this organism produce in terms of "A" and "B" alleles if crossing over does not occur?
    a. *AA* and *BB*
    b. *AB* and *ab*
    c. *Ab* and *aB*
    d. *AB, ab, Ab,* and *aB*
    e. *aa* and *BB*

13.   What gametes can this organism produce in terms of "A" and "B" alleles if crossing over occurs?
   a.   *AA* and *BB*
   b.   *AB* and *ab*
   c.   *Ab* and *aB*
   d.   *AB, ab, Ab*, and *aB*
   e.   *aa* and *BB*

14.   Assume crossing over occurs. What genotypes may be seen in the offspring from a *AABB* crossed with an *AaBb*?
   a.   *AABB*
   b.   *AaBb*
   c.   *AABB* and *AaBb*
   d.   *AABB, AABb, AaBB*, and *AaBb*
   e.   *AABB, AABb, AAbb, AaBB, AaBb, Aabb, aaBB, aaBb*, and *aabb*

15.   Genes "C" and "D" are found on the same chromosome. Allele *C* and *d* are on one homologue and *c* and *D* on the other. Alleles *c* and *d* are found to occur together in the same individual about 300 out of every 1500 offspring. How many map units apart are genes "C" and "D?"
   a.   0.2
   b.   2
   c.   5
   d.   20
   e.   200

16.   Genes "E" and "F" are found on the same chromosome. Allele *E* and *f* are on one homologue and *e* and *F* on the other. Alleles *E* and *F* are found to occur together in the same individual about 60 out of every 1000 offspring. How many map units apart are genes "E" and "F?"
   a.   3
   b.   6
   c.   30
   d.   60
   e.   600

17.   Gender differences in chromosome appearance result from a pair of chromosomes known as
   a.   autosomes.
   b.   sex chromosomes.
   c.   linkage groups.
   d.   carriers.
   e.   pedigrees.

18.   Chromosome pairs identical in both sexes are known as
   a.   autosomes.
   b.   sex chromosomes.
   c.   linkage groups.
   d.   carriers.
   e.   pedigrees.

19.   The somatic cells of human females contain
   a.   11 pairs of autosomes and two X chromosomes.
   b.   11 pairs of sex chromosomes and two X chromosomes.
   c.   11 pairs of autosomes, a Y chromosome, and an X chromosome.
   d.   22 pairs of autosomes and two X chromosomes.
   e.   22 pairs of sex chromosomes and two X chromosomes.

20. The somatic cells of human males contain
   a. 11 pairs of autosomes and 2 X chromosomes.
   b. 11 pairs of sex chromosomes and 2 X chromosomes.
   c. 22 pairs of autosomes, a Y chromosome, and an X chromosome.
   d. 22 pairs of autosomes and 2 X chromosomes.
   e. 22 pairs of sex chromosomes and 2 X chromosomes.

21. Part of the human male phenotype may be determined by a single recessive allele
   a. on an autosome.
   b. on the X chromosome.
   c. in a linkage group.
   d. in a translocated portion of an autosome.
   e. in a carrier.

22. Part of the human male phenotype may be determined by a single recessive allele
   a. on an autosome.
   b. in an inversion.
   c. in a duplication.
   d. on the Y chromosome.
   e. in his germ cells.

23. Chromosomal aberrations include a change in the sequence of genes on a chromosome. This is known as
   a. a carrier.
   b. an inversion.
   c. a duplication.
   d. a translocation.
   e. polyploidy.

24. Chromosomal aberrations include a change in the number of sets of chromosomes. The result may change cells from diploid (2n) to triploid (3n). This is known as
   a. a carrier.
   b. an inversion.
   c. a duplication.
   d. a translocation.
   e. polyploidy.

Concept Map Construction

Construct a concept map for each group of terms. Be sure to include appropriate connector phrases. You may add other terms as necessary.

   1. Law of Independent Assortment, linkage group, X chromosome, gene, map unit
   2. homologous chromosome, mutant, genetic marker, crossing over, testcross
   3. polytene chromosome, chromosomal aberration, mutagenic, translocation, inversion

Chapter 14

The Molecular Basis of Genetics

Section Concept Map

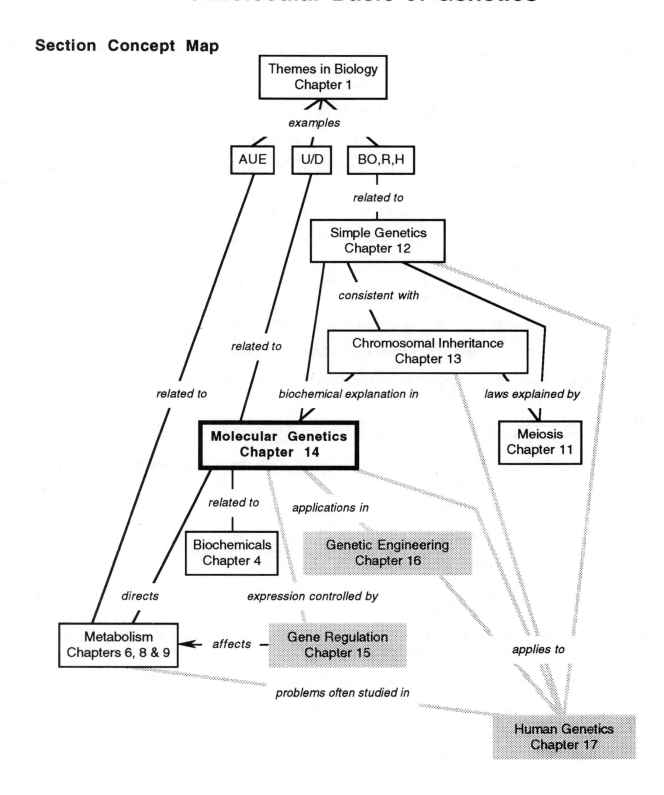

Go Figure!

Figure 14-1: How are the newly formed phage particles able to synthesize protein and DNA without labels when the original particle had either its protein or DNA labeled?

What conclusion may have been drawn if the radioactivity remained outside the bacterial cell?

Figure 14-2: To which sugar carbon atom is the phosphate attached?

How about the nitrogenous base?

Figure 14-3: In what obvious way would the DNA molecule be altered if base pairing involved complementarity between guanine and adenine, and cytosine and thymine? (You may want to look at Figure 14-4 before answering.)

Figure 14-4: Name the bases for each nucleotide in this schematic representation of DNA in this figure.

Figure 14-6: What is accomplished by packaging eukaryotic DNA?

Suggest reasons why this degree of packaging is not seen in prokaryotes.

Figure 14-8: Assume this is eukaryotic DNA. After DNA replication, there are four strands of DNA: two original and two newly synthesized. How are these four strands distributed in the chromatids of a eukaryotic chromosome?

Figure 14-9: Synthesis of a new DNA strand occurs in which direction: 5' to 3' or 3' to 5'?

Why must one strand always be synthesized discontinuously?

Figure 14-10: Since replication occurs in two directions from the origin, is the same template strand copied discontinuously on both sides of the origin? Explain.

Figure 14-12: Since only one strand of DNA (the "sense strand") is necessary for transcription, why is DNA double stranded?

If it had not been indicated for you in this figure, how could you have deduced the direction of transcription?

Figure 14-13: How many different amino acids are shown in the chart?

Which amino acids are encoded by the most codons?

Which amino acid is encoded by the least codons?

Which nucleotide (first, second or third) in a codon appears to be least important in coding for an amino acid?

Figure 14-14: What does the ability of tRNA to base pair with itself tell you about the linear sequence of nucleotides within the molecule?

Figure 14-17: Does the figure illustrate transcription of more than one gene? Justify your answer.

Why are eukaryotes incapable of combining transcription and translation the way prokaryotes do?

Matching Questions

Write the letter of the phrase that best matches the numbered term on the left. Use each only once.

_____ 1.   adenine

_____ 2.   cytosine

_____ 3.   uracil

_____ 4.   tRNA

_____ 5.   rRNA

_____ 6.   mRNA

_____ 7.   replication fork

_____ 8.   semiconservative replication

_____ 9.   nucleosome

_____10.   histone

_____11.   frameshift mutation

_____12.   mutagen

_____13.   transcription

_____14.   translation

_____15.   complementarity

_____16.   DNA polymerase

_____17.   RNA polymerase

_____18.   codon

_____19.   anticodon

_____20.   nonsense codon

a.   three of its nucleotides make up the anticodon

b.   A always pairs with T and G always pairs with C

c.   a unit of DNA packaging

d.   RNA synthesis

e.   occurs as a result of insertion or deletion of a nucleotide

f.   the enzyme of DNA replication

g.   three adjacent nucleotides on tRNA

h.   a purine of DNA

i.   three nucleotides on mRNA for which there is no anticodon

j.   the site of DNA strand separation

k.   a structural component of ribosomes

l.   three adjacent nucleotides on mRNA

m.   the enzyme of transcription

n.   protein associated with DNA

o.   a pyrimidine of DNA

p.   protein synthesis

q.   both strands act as a template during DNA synthesis

r.   it contains the codons

s.   something that changes DNA

t.   the nucleotide that replaces thymine in RNA

Multiple Choice Questions

1. Who was responsible for identifying DNA as the molecule responsible for transformation in bacteria?
    a. Franklin
    b. Avery
    c. Watson and Crick
    d. Hershey and Chase
    e. Griffith

2. Who worked with bacteriophage and determined that DNA is the genetic material?
    a. Franklin
    b. Avery
    c. Watson and Crick
    d. Hershey and Chase
    e. Griffith

3. Which DOES NOT describe DNA?
    a. It forms a double helix.
    b. Adjacent nucleotides are covalently bonded together.
    c. Each nucleotide contains one of four nitrogenous bases.
    d. The sugar in each nucleotide is ribose.
    e. The chains of nucleotides run in "opposite" directions.

4. Which DOES NOT describe DNA?
    a. The nitrogenous bases extend towards the center of the molecule.
    b. Complementary bases are hydrogen bonded to each other.
    c. An adenine nucleotide in one strand base pairs with a thymine nucleotide in the other.
    d. The number of purine nucleotides equals the number of pyrimidine nucleotides in any DNA molecule.
    e. The ends of each strand are identified as 3' or 5'.

5. Which describes the prokaryotic chromosome?
    a. It is a circular molecule of DNA.
    b. It is composed of DNA and histone proteins.
    c. When fully extended, it is shorter than the length of the bacterial cell.
    d. It is organized into nucleosomes.
    e. It is located within the nucleus.

6. Which DOES NOT describe the eukaryotic chromosome?
    a. It is made of a linear DNA molecule.
    b. Each chromosome is associated with histone proteins.
    c. Chromosomal DNA is organized into nucleosomes.
    d. Chromosomes are located in the nucleus.
    e. It is shorter than a prokaryotic chromosome.

7. Which DOES NOT describe DNA replication?
    a. DNA polymerase is responsible for inserting new nucleotides.
    b. Replication is semiconservative.
    c. One new strand undergoes continuous synthesis, whereas the other undergoes discontinuous synthesis.
    d. The replication fork is the site of strand separation.
    e. Replication begins at a single origin for each DNA molecule in eukaryotes.

8. Which DOES NOT describe DNA replication?
   a. Each strand acts as a template for synthesis of a new strand.
   b. There are two replication forks in prokaryotes.
   c. There are RNA molecule responsible for unwinding the DNA helix.
   d. Replication begins at a single origin in prokaryotes.
   e. Prokaryotic DNA polymerase has an active site that allows it to "proofread" the accuracy of nucleotide insertion.

9. What is the complementary strand of DNA to the DNA sequence below?

   3'GTAGCCGTAACGTAT5'

   a. 3'CATCGGCATTGCATA5' $G \rightarrow C$
   b. 3'CAUCGGCAUUGCAUA5' $T \rightarrow A$
   c. 5'CATCGGCATTGCATA3' $C \rightarrow G$
   d. 5'CAUCGGCAUUGCAUA3'
   e. 5'GTAGCCGTAACGTAT3'

10. What is the complementary strand of DNA to the DNA sequence below?

    5'TCGACGGTAGCCTGTG3'

    AGCTGCCATCGGACAC
    a. 3'AGCTGCCATCGGACAC5' $T \rightarrow A$
    b. 3'AGCUGCCAUCGGACAC5' $G \rightarrow C$
    c. 5'AGCTGCCATCGGACAC3' $C \rightarrow G$
    d. 5'AGCUGCCAUCGGACAC3' $A \rightarrow T$
    e. 5'TCGACGGTAGCCTGTG3'

11. What macromolecule is synthesized during transcription?
    a. DNA
    b. RNA
    c. protein
    d. carbohydrate
    e. lipid

12. What macromolecule is synthesized during translation?
    a. DNA
    b. RNA
    c. protein
    d. carbohydrate
    e. lipid

13. Which DOES NOT describe codons?
    a. Each is made of three adjacent nucleotides.
    b. They are part of a tRNA molecule.
    c. They code for amino acid sequence in protein.
    d. Their "meaning" is virtually universal.
    e. There are 64 different codons.

14. Which DOES NOT describe anticodons?
    a. Each is made of three adjacent nucleotides.
    b. They are part of a tRNA molecule.
    c. They base pair with nucleotides in rRNA.
    d. Each corresponds to a particular amino acid.
    e. They are an essential component of translation.

15. Which describes a ribosome?
    a. It consists of two subunits.
    b. It is always in functional form.
    c. It contains a codon.
    d. It contains an anticodon.
    e. It is the site of transcription.

16. Which describes a ribosome?
    a. It is made exclusively of rRNA.
    b. The large subunit contains the enzyme that forms the peptide bond.
    c The small subunit contains the enzyme that forms the peptide bond.
    d. The large subunit binds to the initiation (start) codon (AUG).
    e. The large subunit binds to the initiation (start) anticodon (AUG).

NOTE: Answers for questions 17 through 20 are provided in the Appendix. Questions 21 through 24 constitute a similar sequence, but the answers are not given.

17. What is the mRNA transcribed from the stretch of DNA below?

                   3'TACGGCACGTAT...5'

    a. 3'AUGCCGUGCAUA...5'
    b. 3'ATGCCGUGCAUA...5'
    c. 5'AUGCCGUGCAUA...3'
    d. 5'ATGCCGUGCAUA...3'
    e. 3'UACGGCACGUAU...5'

    T → A
    A → U

18. What is the amino acid sequence coded by the DNA in question #17?
    a. isoleucine-arginine-alanine-valine-...
    b. tyrosine-glycine-threonine-tyrosine-...
    c. methionine-proline-cysteine-isoleucine-...
    d. tyrosine-alanine-arginine-histidine-...
    e. methionine-arginine-serine-isoleucine-...

19. Assume an adenine is inserted between the neighboring guanines of the DNA in question #17. What is the mRNA transcribed from this mutant DNA?
    a. 3'AUGCACGUGCAUA...5'
    b. 3'ATGCACGUGCAUA...5'
    c. 5'AUGCUCGUGCAUA...3'
    d. 5'ATGCACGUGCAUA...3'
    e. 3'UACGUGCACGUAU...5'

20. What is the amino acid sequence coded by the mutant DNA in question #19?
    a. isoleucine-arginine-arginine-serine-...
    b. tyrosine-valine-histidine-isoleucine-...
    c. methionine-proline-cysteine-isoleucine-...
    d. tyrosine-histidine-valine-histidine-...
    e. methionine-leucine-valine-histidine-...

21. What is the mRNA transcribed from the stretch of DNA below?

                   3'TACGAACTAGCG...5'

    a. 5'AUGCUUGAUCGC...3'
    b. 5'ATGCTUGATCGC...3'
    c. 3'UACGAACUAGCG...5'
    d. 3'AUGCUUGAUCGC...5'
    e. 3'ATGCTTGATGCG...5'

22. What is the amino acid sequence coded by the DNA in question #21?
   a. methionine-leucine-aspartic acid-arginine-...
   b. tyrosine-glutamic acid-leucine-alanine-...
   c. tyrosine-glutamic acid-isoleucine-alanine-...
   d. methionine-glutamic acid-leucine-alanine-...
   e. valine-leucine-aspartic acid-arginine-...

23. Assume the adenine in the fifth position of the DNA in question #21 is deleted. What is the mRNA transcribed from this mutant DNA?
   a. 3'UACGACUAGCG...5'
   b. 3'AUGCUGAUCGC...5'
   c. 3'ATGCUGATGCG...5'
   d. 5'ATGCUGATCGC...3'
   e. 5'AUGCUGAUCGC...3'

24. What is the amino acid sequence coded by the mutant DNA in question #23?
   a. tyrosine-glycine-stop
   b. methionine-glycine-stop
   c. methionine-leucine-isoleucine-alanine-...
   d. methionine-proline-isoleucine-...
   e. tyrosine-glycine-isoleucine-stop

Concept Map Construction

Construct a concept map for each group of terms. Be sure to include appropriate connector phrases. You may add other terms as necessary and use terms in the singular or plural form.

1. tRNA, ribosome, semiconservative replication, allele, hypothesis testing
2. protein, Golgi complex, histidine, start codon, homologous chromosome
3. crossing over, frameshift mutation, monohybrid cross, DNA polymerase, metaphase

Chapter 15

Orchestrating Gene Expression

Section Concept Map

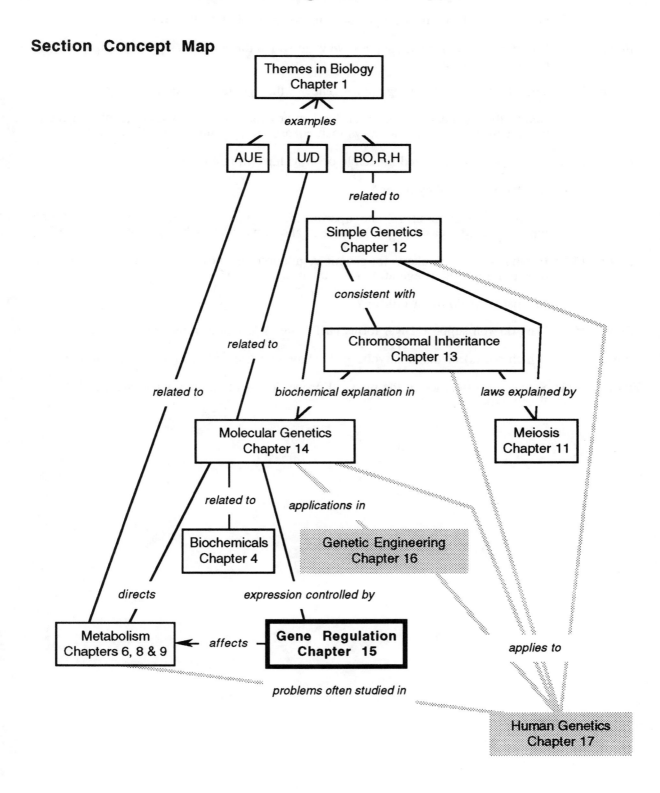

Go Figure!

Figure 15-2: Your text says humans and sea squirts belong to the same phylum. On a genetic level, how do you account for the similarities taxonomists used to place them in the same phylum?

On a genetic level, how do you account for their obvious differences?

Figure 15-3: Based on the illustration, suggest ways in which the regulatory protein might be able to inactivate a gene.

Figure 15-4: What is the role of the regulatory gene? ...the operator? ...the promoter?

Justify the statement: "Bacterial operons provide a more efficient means of genetic control over metabolic pathways than eukaryotic systems of genetic control."

Figure 15-5: What do repressible and inducible operons have in common?

In what ways do they differ?

Justify the statement: "The operon for an anabolic pathway is likely to be repressible, whereas the operon for a catabolic pathway is likely to be inducible."

Figure 15-7: In what ways are eukaryotic genetic regulation (as shown in this figure) similar to prokaryotic genetic regulation (as shown in Figure 15-5)?

In what ways do the two differ?

Figure 15-8: Why do you suppose DNA forms puffs when it is being transcribed?

Figure 15-10: Which part(s) of the eukaryotic gene shown correspond(s) to a prokaryotic gene?

Figure 15-11: Suggest a difference in the physical properties of the amino acids encoded by exon F and exon G.

Matching Questions

Write the letter of the phrase that best matches the numbered term on the left. Use each only once.

_____ 1.   totipotent

_____ 2.   repressible operon

_____ 3.   inducible operon

_____ 4.   inducer

_____ 5.   corepressor

_____ 6.   operator

_____ 7.   promoter

_____ 8.   structural gene

_____ 9.   differentiation

_____10.   exon

_____11.   intron

_____12.   primary transcript

_____13.   ribozyme

_____14.   chromosomal puff

_____15.   ecdysone

a.   DNA that codes for an enzyme

b.   portion of DNA where repressor binds

c.   process of making specialized cells

d.   describes a cell able to give rise to a whole new individual

e.   a coding part of a eukaryotic gene

f.   an insect hormone responsible for metamorphosis

g.   RNA with catalytic activity

h.   binds to repressor and inactivates it

i.   RNA that must be processed to remove introns

j.   genes are not transcribed if substance is present

k.   site of transcription

l.   portion of DNA where RNA polymerase binds

m.   a noncoding part of a eukaryotic gene

n.   genes are transcribed if substance is present

o.   binds to repressor and activates it

Multiple Choice Questions

1. Which DOES NOT describe gene regulatory proteins?
   a. Some are activated as a result of chemical changes in the external environment.
   b. Some are activated as a result of physical changes in the external environment.
   c. Some are activated as a result of hormonal changes in the internal environment.
   d. Some are activated as a result of conscious decisions of the organism.
   e. They are responsible for determining which genes are used in a particular cell at any given time.

2. Which DOES NOT describe gene regulatory proteins?
   a. They are specific for the genes they regulate.
   b. Some are responsible for activating the gene.
   c. Some are responsible for inactivating the gene.
   d. Their precise mechanism of action was easily determined by molecular biologists.
   e. Some are shaped so as to fit the structure of DNA.

3. Which describes an operon?
   a. It is found in prokaryotes.
   b. It consists only of a regulatory gene.
   c. It may only be repressed.
   d. It may only be induced.
   e. It contains a promoter site for producing repressor protein.

4. Which describes an operon?
   a. It is found in eukaryotes.
   b. It contains structural genes for enzymes of a metabolic pathway.
   c. Each structural gene of the operon is regulated independently.
   d. The operator binds RNA polymerase.
   e. The regulatory gene determines whether or not the operon is transcribed.

5. What is the role of the operator in the lac operon?
   a. It binds to RNA polymerase.
   b. It produces the repressor protein.
   c. It produces the inducer.
   d. It binds to the repressor.
   e. It binds to the corepressor.

6. What is the role of the operator in the trp operon?
   a. It binds to RNA polymerase.
   b. It produces the repressor protein.
   c. It produces the inducer.
   d. It binds to the repressor.
   e. It binds to the corepressor.

7. What is the role of the promoter in the lac operon?
   a. It binds to RNA polymerase.
   b. It produces the repressor protein.
   c. It produces the inducer.
   d. It binds to the repressor.
   e. It produces corepressor.

8. What is the role of the promoter in the trp operon?
   a. It binds to RNA polymerase.
   b. It produces the repressor protein.
   c. It produces the inducer.
   d. It binds to the repressor.
   e. It produces corepressor.

9.   What is the role of lactose in the lac operon?
     a.   It is a repressor.
     b.   It is an inducer.
     c.   It is a corepressor.
     d.   It is a substrate for RNA polymerase.
     e.   It is a substrate for DNA polymerase.

10.   What is the role of tryptophan in the trp operon?
      a.   It is a repressor.
      b.   It is an inducer.
      c.   It is a corepressor.
      d.   It is a substrate for RNA polymerase.
      e.   It is a substrate for DNA polymerase.

11.   What happens to an inducible operon when the inducer is present?
      a.   The repressor starts being produced.
      b.   RNA polymerase binds to the inducer.
      c.   The operator is bound to repressor protein.
      d.   Transcription occurs.
      e.   The corepressor is inhibited.

12.   What happens to a repressible operon when the corepressor is present?
      a.   RNA polymerase transcribes the operon.
      b.   The operator binds to the inducer.
      c.   The corepressor binds to the inducer.
      d.   The promoter binds to RNA polymerase.
      e.   The repressor is activated.

13.   What is an intron?
      a.   It is a region of DNA located between genes that is not translated into protein.
      b.   It is a region of DNA located within a gene that is not translated into protein.
      c.   It is a regulatory gene.
      d.   It is a structural gene.
      e.   It is an enzyme.

14.   What is an exon?
      a.   It is a regulatory gene.
      b.   It binds to corepressor.
      c.   It is a segment of a split gene that codes for amino acids in a protein.
      d.   It is a segment of a split gene that has no corresponding amino acids in a protein.
      e.   It is a DNA spacer between genes.

15.   Which describes a eukaryotic transcription-level control mechanism?
      a.   It affects whether or not mRNA is made.
      b.   It regulates an operon.
      c.   It affects protein synthesis by interfering with the ability of ribosomal subunits to assemble.
      d.   It affects gene expression by altering the primary RNA transcript by removing noncoding segments.
      e.   It exerts its control in the cytoplasm.

16.   Which describes a eukaryotic processing-level control mechanism?
      a.   It affects protein synthesis by interfering with the ability of the small ribosomal subunit to bind mRNA.
      b.   It affects the frequency of transcription by interfering with RNA polymerase.
      c.   It affects the frequency of translation by disrupting codon-anticodon interactions.
      d.   It affects the conversion of the primary RNA transcript into functional mRNA that can be translated.
      e.   It affects the longevity of mRNA molecules in the cytoplasm.

17. Which describes ecdysone?
     a. It is a hormone that promotes translation of genes encoding larval proteins.
     b. It is a hormone that promotes transcription of genes encoding proteins necessary for metamorphosis in insects.
     c. It is a hormone found only in male insects that promotes development into an adult male.
     d. It is a hormone found only in female insects that promotes development into an adult female.
     e. It is a cytoplasmic receptor protein that combines with a hormone to initiate development of the larval insect into an adult.

18. Which describes testosterone receptor protein?
     a. The gene for testosterone receptor protein is present in all human cells.
     b. The gene for testosterone receptor protein is present only in some male human cells.
     c. Testosterone receptor protein is bound to membrane.
     d. Testosterone receptor protein is present in all male cells.
     e. Testosterone receptor protein binds to ribosomes to promote translation of the hormone testosterone.

Concept Map Construction

Construct a concept map for each group of terms. Be sure to include appropriate connector phrases. You may add other terms as necessary and use terms in the singular or plural form.

1. transcriptional-level control, chromosomal puff, ribozyme, translation, eukaryote
2. intron, inducible operon, catabolic pathway, DNA, nucleolus
3. interphase, DNA polymerase, microtubule, cell differentiation, homozygote

Chapter 16

DNA Technology: Developments and Applications

Section Concept Map

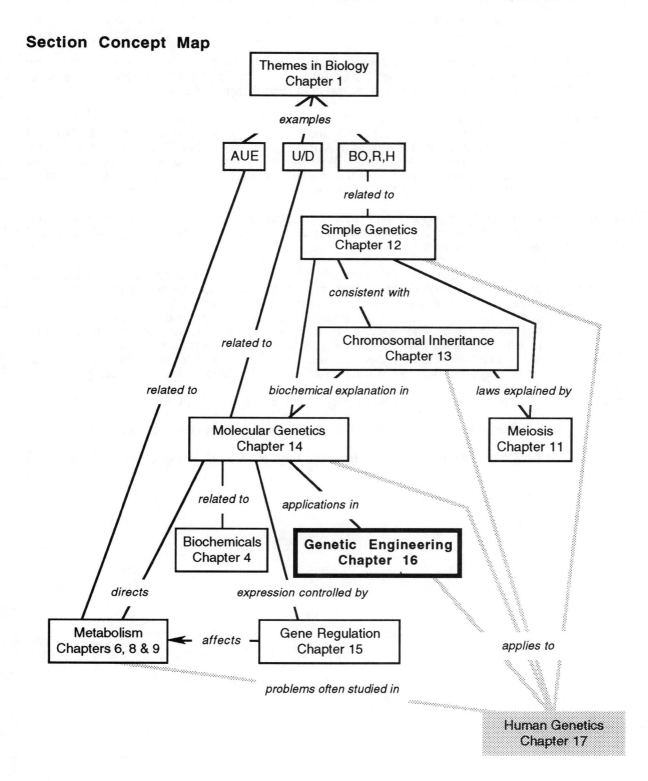

Go Figure!

Figure 16-4: What does the term "transgenic" mean?

How was the rat growth hormone gene introduced into all of the mouse cells?

What problems might be encountered if this procedure was attempted with adult mice?

Figure 16-6: Plasmids are used to introduce foreign genes into bacteria. How might plasmid antibiotic resistance genes be useful to the genetic engineer in determining what bacterial cells have taken in the plasmid (and therefore the foreign gene)?

Figure 16-7: The restriction enzyme Eco R1 always cuts DNA between the T and C of a GAATTC sequence. Other restriction enzymes are Hind III and Bam HI. Hind III cuts between the two T's in an AAGCTT sequence. Bam HI cuts between the two Cs of a GGATCC sequence. Write each of these sequences along with their complementary strands. What pattern is there in the nucleotide sequences "recognized" by these restriction enzymes?

Figure 16-8: Since both the plasmid and the insulin gene (or any gene used in this procedure) have the same sticky ends, what prevents the plasmid from joining with itself and not the insulin gene?

As shown in Figure 16-6, some plasmids have more than one gene for antibiotic resistance. Only bacteria that have taken up the plasmid will grow in an environment containing ampicillin since they now have the ampicillin resistance gene. But this doesn't tell the genetic engineer that the plasmid in the bacterial cell has the desired gene (insulin, in this case). How might the second antibiotic resistance gene be used to help the genetic engineer discriminate between those cells that picked up "empty" plasmids and those that picked up recombinant plasmids? (Hint: suppose there is a restriction site within the tetracycline resistance gene.)

Figure 16-12: Why must the sample be alternately heated and cooled in PCR?

What are the advantages of PCR over cloning using bacterial plasmids?

How might the original DNA fragment used in PCR be obtained and identified?

Matching Questions

Write the letter of the phrase that best matches the numbered term on the left. Use each only once.

_____ 1.   selective breeding

_____ 2.   recombinant DNA

_____ 3.   Frostban

_____ 4.   sense strand

_____ 5.   transgenic animal

_____ 6.   gene therapy

_____ 7.   host cell

_____ 8.   donor cell

_____ 9.   gene splicing

_____10.   vector

_____11.   plasmid

_____12.   restriction enzyme

_____13.   DNA ligase

_____14.   DNA cloning

_____15.   RNA polymerase

_____16.   gel electrophoresis

_____17.   DNA probe

_____18.   polymerase chain reaction

_____19.   DNA fingerprint

a.   this organism contains genes from a different species

b.   this is the transcribed part of a gene

c.   plasmids and viruses are examples

d.   enzyme used to make mRNA from DNA

e.   the cell that receives "foreign" DNA

f.   process used for enzymatic DNA amplification

g.   trade name of ice-minus *Pseudomonas* strain

h.   single stranded DNA used to locate desired restriction fragments after electrophoresis

i.   process used to separate restriction fragments by size

j.   a DNA molecule constructed from pieces derived from different sources

k.   enzyme used to covalently join DNA fragments

l.   allowing only individuals with desirable traits to reproduce

m.   the process of making multiple DNA copies in a bacterial host

n.   small, circular bacterial DNA molecules

o.   gel pattern of unique DNA fragments

p.   a source of DNA used for genetic engineering

q.   replacing a defective gene with a normal one

r.   enzyme that cuts DNA and leaves "sticky ends"

s.   insertion of the desired gene DNA into vector DNA

Multiple Choice Questions

1. Which describes genetic engineering?
    a. It occurs frequently in nature.
    b. It occurs in nature, but not frequently.
    c. It involves combining DNA from individuals of the same species.
    d. It involves the production of recombinant DNA molecules.
    e. It requires enzymes artificially produced in the laboratory.

2. Which describes genetic engineering?
    a. It occurs in the laboratory under human direction.
    b. It occurs in the laboratory, but relies on chance to produce new genotypes.
    c. It relies on mating individuals of the same species to produce new genotypes.
    d. It relies on mating individuals of different species to produce new genotypes.
    e. It has only been successful using bacterial DNA.

3. Which IS NOT a human protein currently produced by genetic engineering?
    a. insulin
    b. tissue plasminogen activator (tPA)
    c. growth hormone
    d. interferon
    e. ethanol

4. Which IS NOT currently produced by genetic engineering?
    a. interleukin 2
    b. hepatitis B vaccine
    c. beta endorphin
    d. ice-nucleating protein
    e. erythropoietin

5. Which organism was the source of the BT toxin engineered into plants as an insecticide?
    a. a bacterium
    b. the gypsy moth
    c. "mushy" tomatoes
    d. a virus
    e. the wheat plant

6. Which describes the transgenic pigs discussed in Chapter 16 of the text?
    a. They contain bacterial genes.
    b. They produce new phenotypes due to mutations induced by irradiation.
    c. They possess extra growth hormone genes.
    d. They are vectors for transmission of plasmids during genetic engineering.
    e. They are being used commercially for treating toxic waste.

7. Which is typically used as a vector in genetic engineering?
    a. bacterial chromosome
    b. bacterial plasmid
    c. human chromosome
    d. bacterial RNA
    e. plant chromosome

8. Which is typically used as a vector in genetic engineering?
    a. animal chromosome
    b. viral DNA
    c. viral RNA
    d. viral plasmid
    e. antisense strand

9. Which describes restriction enzymes?
    a. They form covalent bonds between DNA fragments.
    b. They are creations of genetic engineers and do not exist naturally.
    c. Each acts only on specific nucleotide sequences.
    d. They remove introns.
    e. They act on specific codon sequences in mRNA.

10. Which describes restriction enzymes?
    a. They often produce "staggered cuts" in DNA.
    b. Different restriction enzymes are used to cut vector DNA and the DNA with the desired gene.
    c. They remove exons from RNA.
    d. They remove exons from DNA.
    e. They make clones of the desired gene.

11. What is the next step in a genetic engineering project after a DNA fragment with a desired human gene has been produced?
    a. Plasmid DNA must be cut with the same ligase used to cut the human DNA.
    b. Plasmid DNA must be cut with the same endonuclease used to cut the human DNA.
    c. The fragment must be joined to other fragments containing the same gene so many copies can be made.
    d. Plasmid DNA must be joined with viral DNA to make a vector.
    e. The fragment must be introduced into a bacterial cell.

12. What is the next step in a genetic engineering project after a recombinant DNA molecule has been made?
    a. The recombinant DNA molecule must be cut with restriction enzymes.
    b. The recombinant DNA molecule must be cut with ligase.
    c. The recombinant DNA molecule must be cloned.
    d. The recombinant DNA molecule must be separated by electrophoresis.
    e. The recombinant DNA molecule must be used to produce "sticky ends."

13. Which describes a recombinant DNA molecule in a bacterial cell?
    a. It must contain a human gene.
    b. It must be made of plasmid DNA.
    c. It must contain a prokaryotic promoter.
    d. It must contain prokaryotic RNA polymerase.
    e. It must be cut with a restriction enzyme.

14. Which describes a recombinant DNA molecule in a bacterial cell?
    a. It must be introduced into the bacterial cell by a virus.
    b. It must not have any introns.
    c. It must have a eukaryotic operator region.
    d. It must have a eukaryotic DNA polymerase.
    e. It must have been cloned prior to entry into the bacterial cell.

15. What is the function of gel electrophoresis?
    a. It is used to produce restriction fragments.
    b. It is used to separate restriction fragments.
    c. It is used to clone recombinant DNA molecules.
    d. It is used to repair fragmented recombinant DNA molecules.
    e. It is used to determine the function of the cloned gene.

16. What is the function of the radioactive probe used after gel electrophoresis?
    a. It is used to locate a DNA fragment containing a specific gene.
    b. It is used to join DNA fragments together.
    c. It is used to produce restriction fragments.
    d. It is used to clone recombinant DNA molecules.
    e. It is used to determine the nucleotide sequence of a restriction fragment.

17.   Which describes the polymerase chain reaction (PCR)?
   a.   RNA polymerase from hot springs bacteria is used.
   b.   Protein primers must be used.
   c.   It is a form of cloning.
   d.   The reaction must occur within a bacterial cell.
   e.   Strand separation is accomplished by an enzyme.

18.   Which describes the polymerase chain reaction (PCR)?
   a.   PCR requires large starting amounts of DNA.
   b.   Only synthetic DNA molecules can be used in PCR.
   c.   Only natural DNA molecules can be used in PCR.
   d.   The DNA is alternately heated, then cooled during PCR.
   e.   PCR occurs rapidly, even in the absence of enzymes.

Concept Map Construction

Construct a concept map for each group of terms. Be sure to include appropriate connector phrases.
You may add other terms as necessary and use terms in the singular or plural form.

   1.   polymerase chain reaction, DNA cloning, vector, DNA polymerase, anabolic reaction
   2.   restriction enzyme, plasmid, intron, sticky end, protein
   3.   DNA ligase, DNA probe, restriction fragment, promoter site, dominant allele

Chapter 17

Human Genetics: Past, Present, and Future

Section Concept Map

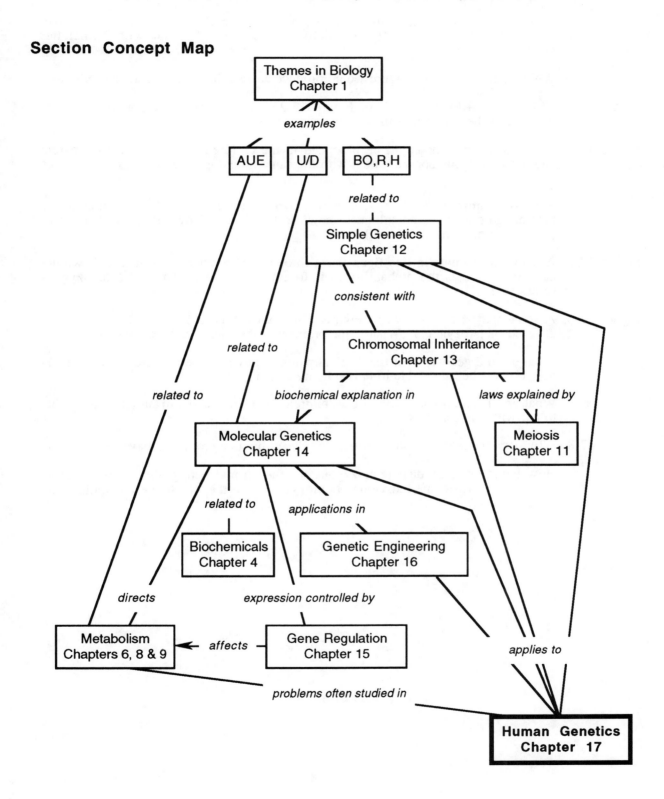

Go Figure!

Figure 17-1: Assume the chromosome involved in nondisjunction is chromosome #21. Which gamete(s) in this figure would produce Down syndrome individuals when combined with a normal gamete at fertilization?

Now assume the chromosome involved in nondisjunction is the X chromosome. Which gamete(s) in this figure would produce an individual with Klinefelter syndrome?

Which gamete(s) in this figure would produce an individual with Turner syndrome?

Figure 17-2: Observe the RFLP patterns in Figure 17-2c. How many different patterns are possible in the children of these parents?

Figure 17-4: What RFLP patterns could be produced in the children of a heterozygous parent and a homozygous normal parent with the same homologues as those shown in this figure?

Suppose another heterozygote with two different homologues marries a heterozygote with the homologues shown in the figure. How many RFLP patterns could be produced in their children?

Figure 17-5: Assume the parents in this figure have three children: one is normal and two are carriers. What is the probability their fourth child will be affected (homozygous recessive)?

Figure 17-6: Assume the parents shown in Figure 17-6a have three children, all of whom are affected (heterozygous). What is the probability the fourth child will be normal?

What are the genotypes of the original parents shown in the pedigree in Figure 17-6b? (It may be impossible to tell with any certainty.)

What are the genotypes of their six children? (In some cases, it may be impossible to tell with any certainty.)

Figure 17-8: Why does the electrophoresis pattern produced by a heterozygote have three bands?

How come an individual homozygous for the normal beta-globin gene produces two bands, while an individual homozygous for the sickle cell anemia gene produces one?

Matching Questions

Write the letter of the phrase that best matches the numbered term on the left. Use each only once.

_____ 1.   gene therapy

_____ 2.   meiotic nondisjunction

_____ 3.   traditional gene mapping

_____ 4.   genetic disorder

_____ 5.   carrier of a gene

_____ 6.   Klinefelter syndrome

_____ 7.   Turner syndrome

_____ 8.   Down syndrome

_____ 9.   Human Genome Project

_____10.   polymorphic gene

_____11.   autosomal recessive allele

_____12.   autosomal dominant allele

_____13.   X-linked recessive allele

_____14.   amniocentesis

_____15.   chorionic villus sampling

a. defective gene or chromosome

b. a gene with several alleles

c. a single allele in males produces the phenotype, whereas two are necessary in females

d. trisomy 21

e. XXY karyotype

f. carried on nonsex chromosome, but must be homozygous to produce its phenotype

g. procedure in which placental tissue is sampled

h. locating a gene's position on a chromosome

i. XO karyotype

j. replacement of a faulty gene

k. procedure in which fetal cells are isolated and cultured

l. heterozygote, if autosomal recessive

m. carried on nonsex chromosome, but produces its phenotype with just one allele present

n. failure of homologs or chromatids to separate

o. attempt to sequence all the human chromosomes

Multiple Choice Questions

NOTE: Answers for questions 1 through 3 are provided in the Appendix. Questions 4 through 6 constitute a similar sequence, but the answers are not given.

Use the following information for questions 1 through 3.

Huntington's disease (HD) is inherited as an autosomal dominant. Patients first exhibit symptoms as adults (90% between the ages of 25 and 60, with the average in the early 40s). They suffer a deterioration of the central nervous system and lose mental abilities and motor control of the limbs. The disease is fatal within about 10 years after the onset of symptoms.

1. What is the probability two individuals heterozygous for HD will have a child with HD?
   a. 0%
   b. 25%
   c. 50%
   d. 75%
   e. 100%

2. What is the probability two individuals homozygous for HD will have a child with HD?
   a. 0%
   b. 25%
   c. 50%
   d. 75%
   e. 100%

3. What is the probability a man with HD but whose mother was normal, and a normal woman will have a child with HD?
   a. 0%
   b. 25%
   c. 50%
   d. 75%
   e. 100%

Use the following information for questions 4 through 6.

An autosomal dominant trait exhibited by some Caucasians is called "wooly hair." The allele responsible for this trait is rare, but when present, produces extremely brittle hair that breaks off before it grows very long.

4. What is the probability of two individuals heterozygous for this trait having a normal child?
   a. 0%
   b. 25%
   c. 50%
   d. 75%
   e. 100%

5. What is the probability of a homozygous wooly haired woman and a heterozygous wooly haired man having a normal child?
   a. 0%
   b. 25%
   c. 50%
   d. 75%
   e. 100%

6. Liz and Dwight are getting married! Liz has wooly hair. Her father also has wooly hair, but her mother has normal hair. Dwight has normal hair, but has a child from another marriage with wooly hair. What is the probability that Liz and Dwight will have a normal child?
   a. 0%
   b. 25%
   c. 50%
   d. 75%
   e. 100%

NOTE: Answers for questions 7 through 9 are provided in the Appendix. Questions 10 through 12 constitute a similar sequence, but the answers are not given.

Use the following information for questions 7 through 9.

Tay-Sachs disease is caused by an autosomal recessive allele that results in deficiency of the enzyme hexoaminidase A. Symptoms include blindness and retardation. Onset of symptoms begins at about six months of age, and death results in early childhood. Genetic counseling of at-risk populations (Ashkenazic Jews and French Canadians in Quebec) has reduced the frequency of the disease.

7. What is the probability that a homozygous normal man and a carrier female will have Tay-Sachs child?
    a. 0%   d. 75%
    b. 25%   e. 100%
    c. 50%

8. What is the probability that the parents in question #7 will have a child who is a Tay-Sachs carrier?
    a. 0%   d. 75%
    b. 25%   e. 100%
    c. 50%

9. Alicia's parents are both normal, but she had a sister who died of Tay-Sachs. What is the probability that Alicia is a carrier of the Tay-Sachs allele?
    a. 0%   d. 75%
    b. 25%   e. 100%
    c. 50%

Use the following information for questions 10 through 12.

Ellis-van Creveld syndrome is a type of dwarfism inherited as an autosomal recessive disease. It is rare in the general population of the United States, but it is unusually common in the Amish people of Pennsylvania. People with this disease have a normal body length, but have shortened limbs and appendages as well as extra fingers or toes.

10. What is the probability that two heterozygotes will have a child that suffers from Ellis-van Creveld syndrome?
    a. 0%   d. 75%
    b. 25%   e. 100%
    c. 50%

11. What is the probability that an Ellis-van Creveld carrier and a normal homozygous individual will have a carrier child?
    a. 0%   d. 75%
    b. 25%   e. 100%
    c. 50%

12. Dick suffers from Ellis-van Creveld syndrome. Margie has a normal stature, but her father suffered from Ellis-van Creveld syndrome. What is the probability that Dick and Margie will produce a dwarf child?
    a. 0%   d. 75%
    b. 25%   e. 100%
    c. 50%

NOTE: Answers for questions 13 through 16 are provided in the Appendix. Questions 17 through 20 constitute a similar sequence, but the answers are not given.

Use the following information for questions 13 through 16.

Lesch-Nyhan syndrome is produced by a recessive allele carried on the X chromosome. Symptoms include compulsive chewing of the lips, fingers, and toes, as well as other abnormal activities. Since this self-destructive behavior is eventually lethal, no females with Lesch-Nyhan syndrome have been identified.

13.   What is the probability of a carrier and a normal male having a carrier child?
  a. 0%   d. 75%
  b. 25%   e. 100%
  c. 50%

14.   What is the probability the sons in question #13 will have Lesch-Nyhan syndrome?
  a. 0%   d. 75%
  b. 25%   e. 100%
  c. 50%

15.   Nathan is a normal teenager. However, he had a brother who died of Lesch-Nyhan syndrome. What is the probability that Nathan is a carrier of the allele for this disease?
  a. 0%   d. 75%
  b. 25%   e. 100%
  c. 50%

16.   Refer to question #15. What are the genotypes of Nathan's parents? (X^* = Lesch-Nyhan allele.)
  a. XY and XX   d. X^*Y and X^*X^*
  b. X^*Y and XX   e. XY and X^*X
  c. XY and X^*X^*

Use the following information for questions 17 through 20.

Hemophilia A is a disease in which patients lack a clotting factor in the blood (factor VIII). It is inherited as an X-linked recessive trait.

17.   What is the probability of a hemophilic male and a carrier female producing a hemophilic daughter?
  a. 0%   d. 75%
  b. 25%   e. 100%
  c. 50%

18.   What is the probability that the parents in question #17 will produce a hemophilic son?
  a. 0%   d. 75%
  b. 25%   e. 100%
  c. 50%

19.   Eric has normal blood. His wife Jane is the daughter of a hemophilic father and a homozygous normal mother. What are the chances that Eric and Jane will have a daughter who is a carrier?
  a. 0%   d. 75%
  b. 25%   e. 100%
  c. 50%

20.   Ralph has normal blood, but he has two brothers and one sister who are hemophilic. What are the genotypes of Ralph's parents? (X^H = normal; X^h = hemophilia)
   a. X^HY and X^HX^H
   b. X^hY and X^HX^H
   c. X^HY and X^hX^h
   d. X^hY and X^hX^h
   e. X^hY and X^HX^h

Questions 21 through 24 use a pedigree to show the inheritance pattern of particular traits in a family. A key to the standard symbols used in constructing pedigrees is provided below and applies to all problems involving pedigrees.

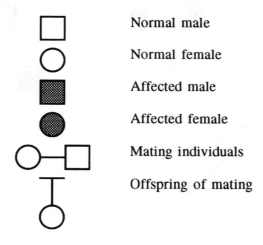

Normal male

Normal female

Affected male

Affected female

Mating individuals

Offspring of mating

21.   Choose the mode of inheritance that best fits the hypothetical pedigree.

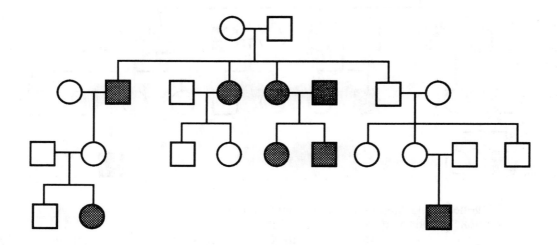

   a. The gene is inherited as an X-linked recessive.
   b. The gene is inherited as an X-linked dominant.
   c. The gene is inherited as an autosomal dominant.
   d. The gene is inherited as an autosomal recessive.
   e. The gene is inherited as a Y-linked dominant.

22. Choose the mode of inheritance that best fits the hypothetical pedigree.

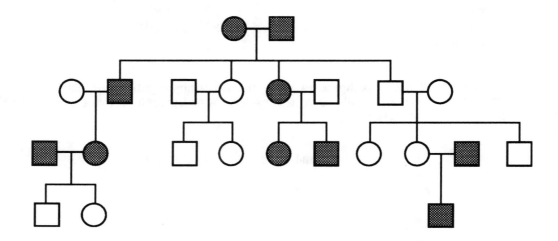

   a. The gene is inherited as an X-linked recessive.
   b. The gene is inherited as an X-linked dominant.
   c. The gene is inherited as an autosomal dominant.
   d. The gene is inherited as an autosomal recessive.
   e. The gene is inherited as a Y-linked dominant.

23. Choose the mode of inheritance that best fits the hypothetical pedigree.

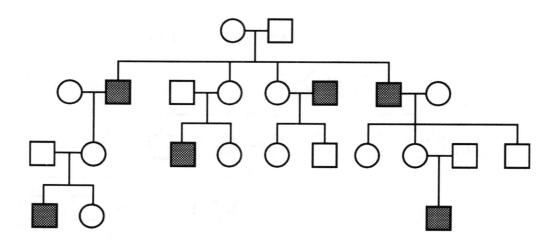

   a. The gene is inherited as an autosomal dominant.
   b. The gene is inherited as an autosomal recessive.
   c. The gene is probably inherited as an X-linked recessive, but autosomal recessive can't be ruled out.
   d. The gene is probably inherited as an X-linked dominant, but autosomal recessive can't be ruled out.
   e. The gene is probably inherited as a Y-linked dominant, but autosomal dominant can't be ruled out.

24.   Choose the mode of inheritance that best fits the hypothetical pedigree.

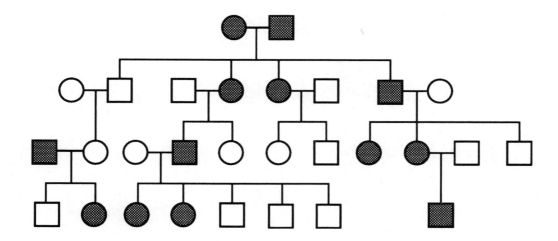

   a.   The gene is inherited as an autosomal dominant.
   b.   The gene is inherited as an autosomal recessive.
   c.   The gene is probably inherited as an X-linked recessive, but autosomal recessive can't be
        ruled out.
   d.   The gene is probably inherited as an X-linked dominant, but autosomal recessive can't be
        ruled out.
   e.   The gene is probably inherited as a Y-linked dominant, but autosomal dominant can't be
        ruled out.

Concept Map Construction

Construct a concept map for each group of terms. Be sure to include appropriate connector phrases.
You may add other terms as necessary and use terms in the singular or plural form.

   1.   nondisjuntion, Turner syndrome, polymorphic gene, X-linked recessive allele, carrier female
   2.   genetic marker, chorionic villus sampling, autosomal recessive allele, amniocentesis, DNA
   3.   restriction enzyme, restriction fragment length polymorphism, gene therapy, genetic disorder,
        gene mapping

Chapter 18

Plant Tissues and Organs

Section Concept Map

Go Figure!

Figure 18-2: In what way are the root and shoot systems related to the dermal tissue, ground tissue, and vascular tissue systems?

Figure 18-3: In what way are apical meristems similar to secondary (lateral) meristems?

How do they differ?

Figure 18-4: The stem in this illustration was cut in cross section. In what way must the stem be cut in order to see what the cells look like lengthwise?

Why are the vascular tissues not discussed in this figure?

Figure 18-5: To what tissue do sclereids belong?

Figure 18-6: What type of tissue comprises the bulk of the ground tissue system?

Figure 18-8: How do guard cells enable a plant to maintain homeostasis during water stress?

Figure 18-10: In what ways are lenticels and stomates similar?

How are they different?

Figure 18-14: Why doesn't secondary phloem accumulate to the same degree as secondary xylem?

Which tissues are included in bark?

Which tissue is "bark" to someone not trained in botany?

Figure 18-16: Compare this figure to Figure 18-20. How do root tips differ from stem tips?

Figure 18-18: Which plant group is characterized by a tap root system?

Which plant group is characterized by a fibrous (adventitious) root system?

Figure 18-21: Which tissue system gives rise to the pericycle?

Why do you suppose lateral roots arise from within the root rather than from the surface?

Figure 18-22: What anatomical structure allows the botanist to distinguish between a stem with many small leaves and a compound leaf with many leaflets (which might be misinterpreted as a stem with many small leaves)?

Figure 18-23: Notice that stem and root xylem is towards the center and phloem is towards the surface (see Figures 18-17 and 18-21). Propose an anatomical reason why leaf xylem is towards the upper leaf surface and phloem is towards the lower leaf surface.

Matching Questions

Write the letter of the phrase that best matches the numbered term on the left. Use each only once.

_____ 1. biennial plant

_____ 2. monocot

_____ 3. dicot

_____ 4. cuticle

_____ 5. root system

_____ 6. shoot system

_____ 7. vascular cambium

_____ 8. apical meristem

_____ 9. vessel member

_____10. sieve-tube member

_____11. pith

_____12. Casparian strip

_____13. parenchyma

_____14. collenchyma

_____15. axillary bud

_____16. compound leaf

_____17. petiole

_____18. pericycle

_____19. vascular tissue system

_____20. dermal tissue system

a. characterized by parallel veins in leaves

b. conducting cell of xylem

c. the stalk of a leaf

d. constitutes the cellular covering of a plant

e. conducting cell of phloem

f. root cells responsible for producing lateral roots

g. the most abundant tissue in herbaceous plants

h. has a lifespan of two years

i. region at the center of a stem or root

j. characterized by flower parts in fours or fives

k. waxy covering on the epidermis

l. located at the junction of each leaf and the stem

m. responsible for secondary growth in a plant

n. composed of many leaflets

o. consists of stems, leaves and flowers

p. found in the walls of endodermal cells

q. support tissue of the primary plant body

r. functions include anchoring the plant and absorbing water

s. region of a plant responsible for primary growth

t. includes all the conducting cells of a plant

Multiple Choice Questions

1. Which kind of plant completes its life cycle in a single year?
    a. annual
    b. biennial
    c. perennial
    d. vascular
    e. nonvascular

2. Which kind of plant lives longer than two years?
    a. annual
    b. biennial
    c. perennial
    d. vascular
    e. nonvascular

3. Which term is applied to the underground portion of a plant?
    a. ground system
    b. dermal system
    c. vascular system
    d. root system
    e. shoot system

4. Which term is applied to the aerial portion of a plant?
    a. ground system
    b. dermal system
    c. vascular system
    d. root system
    e. shoot system

5. Which DOES NOT describe monocots?
    a. parallel leaf venation
    b. flower parts in 4s and 5s
    c. scattered vascular bundles in stem
    d. fibrous root system
    e. no secondary growth

6. Which DOES NOT describe dicots?
    a. two types of photosynthetic cells in leaves
    b. tap root system
    c. pith in center of root
    d. two cotyledons in embryo
    e. stem vascular bundles in rings

7. Which DOES NOT describe primary growth?
    a. It produces primary tissues.
    b. It causes elongation of the plant.
    c. It involves an apical meristem.
    d. It occurs only in annual plants.
    e. It produces vascular and other tissues.

8. Which DOES NOT describe secondary growth?
    a. It produces secondary tissues.
    b. It involves vascular cambium.
    c. It involves cork cambium.
    d. It increases plant diameter.
    e. It produces only simple tissues.

9. Which DOES NOT describe parenchyma?
   a. It is a complex tissue.
   b. It is a common tissue in herbaceous plants.
   c. Parenchyma cells are versatile in structure and function.
   d. It is often involved in photosynthesis.
   e. It is found on the surface of the primary plant body.

10. Which DOES NOT describe collenchyma?
    a. The cells are isodiametric.
    b. The cells often have unevenly thickened cell walls.
    c. It physically supports growing parts of the plant.
    d. The cells perform photosynthesis.
    e. It lies just below the epidermis in stems and leaves.

11. Which describes sclerenchyma fibers?
    a. They constitute a complex tissue.
    b. They have thin primary and secondary walls.
    c. They are long and thin.
    d. The secondary wall is soft to allow growth.
    e. The cells are only functional when living.

12. Which describes sclereids?
    a. They are a type of collenchyma cell.
    b. They have hard, thick secondary walls.
    c. They typically are long and thin.
    d. They come in only one shape.
    e. The cells are only functional when living.

13. Which DOES NOT describe epidermis?
    a. It is the cellular covering of the primary plant body.
    b. The epidermal cells of the shoot system are covered with a waxy cuticle.
    c. It contains openings to the interior of the plant called stomates.
    d. It belongs to the dermal tissue system.
    e. Its primary function is transport of material between parts of the plant.

14. Which DOES NOT describe peridermis?
    a. It contains openings to the interior of the plant called stomates.
    b. It is the cellular covering of the secondary plant body.
    c. It is a component of bark.
    d. It belongs to the dermal tissue system.
    e. It contains suberized cork cells.

15. Which DOES NOT describe xylem?
    a. It belongs to the vascular tissue system.
    b. The conducting cells are functional when dead.
    c. It transports water and minerals.
    d. Secondary xylem is also called bark.
    e. Vessel members and tracheids are the conducting cells of xylem.

16. Which DOES NOT describe phloem?
    a. The conducting cells are functional when dead.
    b. The conducting cells are called sieve-tube members.
    c. It transports photosynthetic products.
    d. It belongs to the vascular tissue system.
    e. The conducting cells are located next to companion cells.

17. Which belongs to the ground tissue system?
    a. vascular cylinder
    b. epidermis
    c. cortex
    d. cork
    e. pericycle

18. Which belongs to the ground tissue system?
    a. pith
    b. xylem
    c. phloem
    d. companion cells
    e. guard cells

19. Which DOES NOT describe a primary stem?
    a. There is a meristem at the tip.
    b. There is an active cambium.
    c. Some growth is due to lengthening of cells.
    d. It may belong to a dicot.
    e. It may belong to a monocot.

20. Which DOES NOT describe a plant in secondary growth?
    a. There is an active cambium.
    b. It may be a monocot.
    c. It may be a dicot.
    d. There is an active meristem at the tips of each shoot.
    e. Its secondary xylem is also known as wood.

21. Roots arising from stem tissue are called
    a. storage roots.
    b. tap roots.
    c. root hairs.
    d. adventitious roots.
    e. lateral roots.

22. Where are root hairs located?
    a. on the root cap
    b. at the apical meristem
    c. at the region of elongation
    d. at the region of maturation
    e. at the Casparian strip

23. What is the function of root hairs?
    a. They increase surface area for water and mineral absorption.
    b. They increase surface area for anchorage of the plant.
    c. They increase volume of the root for storage.
    d. They add cells to the root meristem.
    e. They form a waterproof barrier on the surface of the root.

24. What is the function of endodermis?
    a. It insures that water and minerals only enter the vascular cylinder through living cells.
    b. It increases surface area for absorption.
    c. It gives rise to lateral roots.
    d. It acts as a storage tissue.
    e. It performs photosynthesis.

25. What is the flat part of a leaf called?
    a. node
    b. internode
    c. petiole
    d. blade
    e. axillary bud

26. Which tissue is found in the central portion of a leaf?
    a. epidermis
    b. mesophyll
    c. endodermis
    d. guard cells
    e. peridermis

Concept Map Construction

Construct a concept map for each group of terms. Be sure to include appropriate connector phrases. You may add other terms as necessary and use terms in the singular or plural form.

1. vascular plant, monocot, primary growth, companion cell, annual plant
2. lenticel, epidermis, ground tissue system, mesophyll, Photosystem I
3. growth ring, tracheid, meristem, Casparian strip, lateral root

Chapter 19

The Living Plant: Circulation and Transport

Section Concept Map

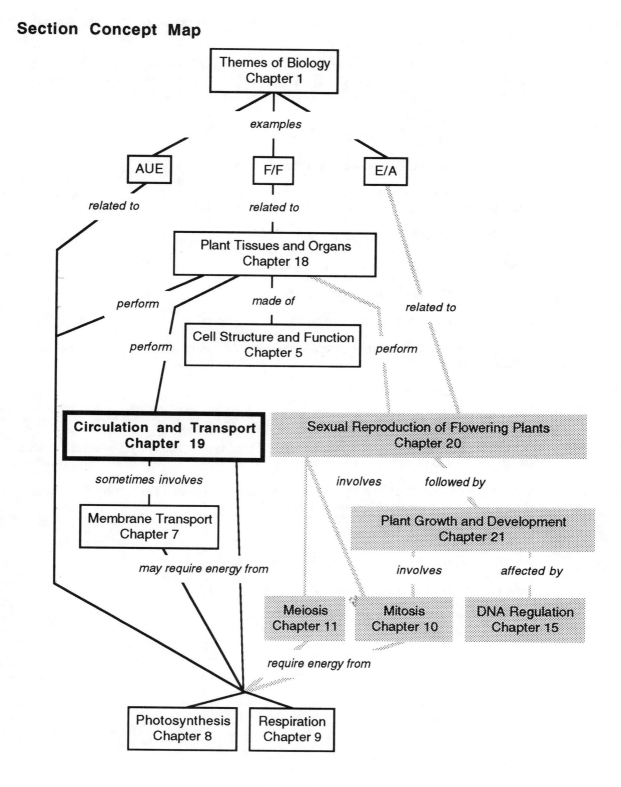

Go Figure!

Figure 19-1: What physical characteristic makes fungi efficient water absorbers?

Suggest reasons why 100% of the species in the North Carolina forest study didn't show evidence of mycorrhizal linkages.

Figure 19-2: Would water enter roots as efficiently if minerals were extracted from the air rather than the soil? Defend your answer.

Observe the endodermal cells. Each cell has six walls: one on each side, an upper and lower, and a front and back. The figure shows the Casparian strip in the two side walls. Which other walls, if any, must also contain a Casparian strip to produce an effective barrier?

What purpose does the Casparian strip serve?

What are the connections between cortical cells called?

Figure 19-3: Nitrogen gas (N_2) comprises almost 80% of air. Why are nitrogen-fixing bacteria essential to the survival of plants in nitrogen-poor soils?

Figure 19-4: Why does guttation only occur at night?

Figure 19-5: What are the conducting cells of xylem called?

Are they alive or dead when functional?

What structure do they form when lined up end to end?

How do growth and metabolism contribute to the pressure gradient necessary for xylem transport?

Figure 19-6: What cell layers are shown in the leaf diagram?

What is the opening in the lower leaf surface called?

Why is most water lost through openings of this type when they constitute such a small percentage of leaf surface area?

Figure 19-7: Since 90% of the water lost by a plant exits through the stomates, why don't plants keep them closed all the time and minimize their loss?

Figure 19-8: Why are aphids considered pests by many plant growers?

Figure 19-9: Are the tissues at the top of the figure acting as sources or sinks?

How can storage tissue act both as a source and a sink?

Matching Questions

Write the letter of the phrase that best matches the numbered term on the left. Use each only once.

_____ 1. xylem

_____ 2. phloem

_____ 3. sink

_____ 4. source

_____ 5. stomates

_____ 6. transpiration

_____ 7. root nodule

_____ 8. mycorrhizae

_____ 9. micronutrient

_____ 10. macronutrient

_____ 11. osmosis

_____ 12. phloem loading

_____ 13. phloem unloading

_____ 14. plasmodesmata

_____ 15. root pressure

_____ 16. guttation

a. process of evaporative water loss

b. process responsible for water entering a root

c. element required only in small amounts for normal plant growth

d. tissue responsible for organic molecule transport

e. process of water loss due only to root pressure

f. occurs at the sink

g. tissue responsible for mineral and water transport

h. produced by an influx of water into roots

i. connections between the cytoplasm of plant cells

j. fungi associated with roots of most vascular plants

k. element necessary in large quantities for normal plant growth

l. growing plant parts or storage tissues

m. primary site of water loss by a plant

n. occupied by nitrogen-fixing bacteria

o. occurs at the source

p. photosynthetic or storage tissues

Multiple Choice Questions

1. Which tissue is responsible for water and mineral transport up a vascular plant?
   a. parenchyma
   b. epidermis
   c. peridermis
   d. xylem
   e. phloem

2. Which tissue is responsible for organic compound transport throughout a vascular plant?
   a. parenchyma
   b. epidermis
   c. peridermis
   d. xylem
   e. phloem

3. Which DOES NOT describe mycorrhizae?
   a. They involve vascular plants and fungi.
   b. The fungus increases absorbing surface area for the plant.
   c. They make plants more efficient at absorbing light energy.
   d. They may be essential to the plant's survival.
   e. They may not be essential to the plant's survival.

4. Which DOES NOT describe mycorrhizae?
   a. Only the fungus benefits from the "partnership."
   b. They may result in connections between plants.
   c. The fungus obtains food from the plant.
   d. They are common in vascular plants.
   e. Roots are the plant organ involved.

5. Which describes the process of water absorption by vascular plants?
   a. Water is actively transported into roots.
   b. Water moves only through living cells in getting to the vascular tissue.
   c. Water movement is along its concentration gradient.
   d. Water movement is independent of mineral movement.
   e. Water movement is restricted in roots by the cuticle.

6. Which describes the process of mineral absorption by vascular plants?
   a. Minerals are actively transported into roots.
   b. Minerals are destined for phloem tissue in the root.
   c. Minerals typically move along their concentration gradient.
   d. Minerals are concentrated in epidermal cells to produce a concentration gradient favorable to mineral movement into the xylem.
   e. Plasmodesmata are too small to allow passage of macronutrients.

7. Which is classified as a macronutrient?
   a. chlorine (Cl)
   b. calcium (Ca)
   c. copper (Cu)
   d. manganese (Mn)
   e. zinc (Zn)

8. Which is classified as a micronutrient?
   a. carbon (C)
   b. nitrogen (N)
   c. phosphorus (P)
   d. sulfur (S)
   e. iron (Fe)

9. Which is a function of potassium (K)?
    a. It is a major component of organic molecules.
    b. It helps regulate stomatal opening and closing.
    c. It acts as an enzyme cofactor.
    d. It acts as a coenzyme.
    e. It is an electron acceptor in photosynthesis.

10. Which is a function of iron (Fe)?
    a. It is a component of cytochromes.
    b. It is a component of the middle lamella.
    c. It helps regulate stomatal opening and closing.
    d. It is a major component of ATP.
    e. It a component of Coenzyme A.

11. Which describes nitrogen fixation?
    a. Most is performed by plants.
    b. It converts nitrogen gas (N_2) to protein.
    c. It converts ammonia (NH_3) to nitrogen gas (N_2).
    d. It converts nitrogen into a form usable by plants.
    e. It leaves soils deficient in other essential nutrients.

12. Which describes nitrogen fixation?
    a. Most is performed by mycorrhizae.
    b. It may be performed by nitrogen-fixing bacteria in the soil.
    c. It converts nitrate (NO_3) into ammonia (NH_3).
    d. It converts nitrate (NO_3) into nitrogen gas (N_2).
    e. It is not an essential process since atmospheric nitrogen is abundant.

13. Which describes xylem transport?
    a. Most materials are pushed up the stem by root pressure.
    b. Materials are moved primarily by guttation.
    c. Xylem transport occurs at a uniform rate throughout the day.
    d. Materials in the xylem are under tension.
    e. Xylem transport requires cellular active transport.

14. Which describes xylem transport?
    a. Evaporation of water from leaves results in movement up the xylem.
    b. Pressure from carbohydrate absorption pushes materials up the xylem.
    c. It requires facilitated diffusion of minerals and water follows by osmosis.
    d. It involves pressure produced from breaks in the water column in xylem vessels.
    e. Loading of xylem vessels is by endocytosis.

15. Which results in faster transpiration?
    a. increased humidity
    b. increased temperature
    c. decreased soil surface
    d. decreased root surface
    e. decreased pressure within guard cells

16. Which results in faster transpiration?
    a. increased breeze
    b. increased soil surface
    c. increased ATP hydrolysis
    d. decreased leaf surface
    e. decreased light

17. Which DOES NOT describe phloem transport?
    a. Phloem loading occurs at the source.
    b. Movement is due to positive pressure in sieve tubes.
    c. Pressure is higher at the sink than at the source.
    d. ATP hydrolysis is required at the source.
    e. Flow is from source to sink.

18. Which DOES NOT describe phloem transport?
    a. ATP hydrolysis is required at the sink.
    b. Sucrose is the main organic compound transported.
    c. A pressure gradient exists in sieve tubes.
    d. ATP for transport is supplied by sieve-tube members.
    e. Phloem unloading occurs at the sink.

19. How are phloem and xylem transport different?
    a. Xylem transport requires cellular energy; phloem transport does not.
    b. Xylem transport is exclusively up the stem; phloem transport is exclusively down the stem.
    c. Xylem transport occurs in cells that are dead; phloem transport occurs in cells that are alive.
    d. Xylem transport occurs as a result of positive pressure; phloem transport occurs as a result of negative pressure.
    e. Xylem transport occurs only in roots; phloem transport occurs only in leaves.

20. How are phloem and xylem transport different?
    a. Phloem transport requires cellular energy; xylem transport does not.
    b. Phloem transport is exclusively up the stem; xylem transport is exclusively down the stem.
    c. Phloem transport occurs in cells that are dead; xylem transport occurs in cells that are alive.
    d. Phloem transport occurs as a result of suction; xylem transport occurs as a result of pushing.
    e. Phloem transport occurs only in stems; xylem transport occurs only in roots.

Concept Map Construction

Construct a concept map for each group of terms. Be sure to include appropriate connector phrases. You may add other terms as necessary and use terms in the singular or plural form.

1. transpirational pull, stomate, vessel member, humidity, root pressure
2. cortex, root hair, mycorrhizae, active transport, parenchyma
3. ATP, source, companion cell, cohesion, phloem loading

Chapter 20

Sexual Reproduction of Flowering Plants

Section Concept Map

Go Figure!

Figure 20-2: All parts of the flower are thought to be modified leaves. Suggest ways in which hollow structures such as the pistil and anthers could be formed from flat leaves.

Figure 20-5: Why do feathery stigmas and production of large amounts of pollen improve chances of successful pollination in wind pollinated flowers?

Figure 20-6: Monoecious plants and plants with perfect flowers are bisexual and self-pollination (and therefore self-fertilization) is a possibility. Assume self-pollination occurs. Will the egg and sperm be identical since they came from the same individual? Defend your answer based on what you know about meiosis.

Figure 20-7: What process produces megaspores in the plant life cycle?

What process produces egg gametes in the plant life cycle?

What structure represents the female gametophyte

What is the ploidy of the female gametophyte?

How do spores differ from gametes? How are they similar?

Figure 20-8: What process produces microspores in the plant life cycle?

What process produces sperm in the plant life cycle?

What structure represents the male gametophyte?

What is the ploidy of the male gametophyte?

Figure 20-10: Distinguish between the processes of pollination and fertilization.

Figure 20-11: What process makes endosperm tissue multicellular?

How does the single celled zygote give rise to the distinctly different embryo and suspensor?

Besides reading the label, how could you tell this is a diagram of a dicot embryo?

Figure 20-12: Besides reading the label, how could you tell this is a diagram of a monocot embryo?

How many ovules are in each ovary of the corn plant?

Figure 20-14: Most people eat apples with the "stem" on the top. Relative to the flower that produced the fruit, do we eat apples right side up or upside down?

Figure 20-15: What is the advantage of efficient fruit dispersal?

Figure 20-18: Compare the third seedling drawing (second from right) to the final early embryo drawing shown in Figure 20-11. Notice all parts of the seedling are present in the embryo depicted at the lower left (Figure 20-11f). Seedling production simply involves differential growth of the embryonic root and shoot at the expense of the cotyledons. What important function have the cotyledons performed between Figure 20-11f and the first drawing in Figure 20-18?

Why is it incorrect to say the epicotyl is solely responsible for producing the stem of the seedling?

Why do the cotyledons wither soon after germination?

Figure 20-19: Compare the corn seedling drawing to the embryo shown in Figure 20-12. Notice all parts of the seedling are present in the embryo. Germination simply involves growth of the embryo (the corn kernel stays the same size).

What is the role of the corn cotyledon?

How is its role different from dicot cotyledons?

What other difference between monocot and dicot cotyledons is evident in Figures 20-18 and 20-19?

Matching Questions

Write the letter of the phrase that best matches the numbered term on the left. Use each only once.

_____ 1. pollination

_____ 2. sepal

_____ 3. petal

_____ 4. anther

_____ 5. stigma

_____ 6. endosperm cell

_____ 7. nectary

_____ 8. generative cell

_____ 9. dioecious

_____10. mitosis

_____11. meiosis

_____12. megaspore

_____13. imperfect flower

_____14. complete flower

_____15. female gametophyte

_____16. male gametophyte

_____17. ovary

_____18. ovule

_____19. zygote

_____20. germination

a. pollen producing part of the flower

b. a flower that lacks stamens or pistils

c. develops into the embryo

d. site where pollen lands

e. process that produces spores

f. a perfect flower with sepals and petals

g. process of pollen transfer between flowers

h. develops into the embryo sac

i. develops into a fruit

j. cell that produces sperm

k. typically a brightly-colored flower part

l. first stage of seedling development

m. develops into a seed

n. a cell with two haploid nuclei

o. process that produces gametes in plants

p. leaflike structure that protects a developing flower

q. the embryo sac

r. different plants produce male and female flowers

s. the pollen grain

t. a gland with a sugary secretion

Multiple Choice Questions

1. Which is the process of pollen transfer?
   a. germination
   b. fertilization
   c. pollination
   d. parthenocarpy
   e. dormancy

2. Which is the process of gamete fusion?
   a. germination
   b. fertilization
   c. pollination
   d. parthenocarpy
   e. dormancy

3. Which DOES NOT describe flower structure?
   a. Large, showy petals attract pollinators.
   b. The stigma produces the female gametophyte.
   c. All parts of the flower are thought to be modified leaves.
   d. The anthers produce male gametophytes.
   e. The flower parts are attached to the receptacle.

4. Which DOES NOT describe flower structure?
   a. The pistil is made of modified leaves called carpels.
   b. Sepals protect the developing flower.
   c. Stamens are the male part of the flower.
   d. The pistil is the female part of the flower.
   e. Many eggs develop within the embryo sac.

5. *Gossypium hirsutum* is the cotton plant. Its flowers occur singly and are positioned above three leaf-like bracts. The calyx is small and cup-shaped, and composed of five sepals. The rest of the flower consists of a corolla with five petals, numerous stamens forming a tube around the style, and an ovary with three to five carpels, each with several ovules. Which DOES NOT describe *Gossypium hirsutum*?
   a. The flowers are perfect.
   b. The flowers are complete.
   c. *Gossypium hirsutum* is dioecious.
   d. *Gossypium hirsutum* is an angiosperm.
   e. Each fruit should contain many seeds.

6. *Papaver somniferum* is the poppy plant. It is the source of opium and morphine, as well as the poppy seeds used by the baking industry. Its flowers are borne singly and consist of two sepals that are deciduous (they drop off), four petals, numerous stamens, and an ovary made of many fused carpels containing many ovules. Which DOES NOT describe *Papaver somniferum*?
   a. The flowers are perfect.
   b. The flowers are complete.
   c. *Papaver somniferum* is monoecious.
   d. *Papaver somniferum* is an angiosperm.
   e. Each fruit should contain a single seed.

7. *Cucumis melo* is the cantaloupe. Plants are monoecious and the flowers are imperfect. Which describes *Cucumis melo*?
   a. Male and female flowers are on different plants.
   b. Flowers possess both stamens and pistils, but lack either petals or sepals.
   c. Flowers possess both stamens and pistils, but may or may not have both sepals and petals.
   d. Each flower lacks either stamens or pistils.
   e. Male flowers produce fruit with more than one seed.

8. *Beta vulgaris* is the sugar beet plant, grown for its edible root. Its flowers are perfect, but incomplete. Which describes *Beta vulgaris*?
   a. The flowers have both stamens and pistils, but lack either sepals or petals.
   b. The flowers have stamens, pistils, sepals and petals.
   c. The flowers lack either stamens or pistils, but have sepals and petals.
   d. The flowers lack stamens or pistils, and also lack either sepals or petals.
   e. *Beta vulgaris* is dioecious.

9. Which is the immediate product of meiosis in the pistil of a flower?
   a. sperm
   b. egg
   c. microspore
   d. megaspore
   e. embryo sac

10. Which is the immediate product of meiosis in the anther of a flower?
    a. sperm
    b. egg
    c. microspore
    d. megaspore
    e. embryo sac

11. What is another name for the male gametophyte?
    a. embryo sac
    b. anther
    c. endosperm
    d. pollen
    e. microspore

12. What is another name for the female gametophyte?
    a. embryo sac
    b. anther
    c. endosperm
    d. pollen
    e. megaspore

13. Which describes the embryo sac?
    a. It consists of eight cells and eight nuclei.
    b. All the cells are diploid.
    c. The generative cell produces the egg.
    d. It contains the egg.
    e. It develops from the egg.

14. Which describes the embryo sac?
    a. It consists of two cells.
    b. All the cells are haploid.
    c. The tube cell receives the pollen tube during fertilization.
    d. It is contained within the anther.
    e. It develops from a megaspore.

15. Which describes the pollen grain?
    a. It consists of seven cells and eight nuclei.
    b. All the cells are haploid.
    c. It is surrounded by one or two integuments.
    d. The endosperm cell is triploid.
    e. It develops from a megaspore.

16. Which describes the pollen grain?
    a. The tube cell is diploid.
    b. The generative cell is diploid.
    c. It develops into a microspore.
    d. The generative cell produces two sperm cells.
    e. Sperm escape the pollen grain through its micropyle.

17. Which DOES NOT describe a dicot seed?
    a. The embryo has two cotyledons.
    b. Each cotyledon contains endosperm.
    c. The radicle is part of the epicotyl.
    d. It developed from an ovule.
    e. The seed coats developed from the integuments.

18. Which DOES NOT describe a monocot seed?
    a. Each cotyledon contains endosperm.
    b. The embryo has one cotyledon.
    c. It developed from an ovule.
    d. The epicotyl develops into most of the shoot system.
    e. Endosperm is triploid tissue.

19. Each fruit of an orange tree comes from the ripened ovary of a single flower. How is the orange fruit classified?
    a. simple fruit
    b. aggregate fruit
    c. multiple fruit
    d. vegetative fruit
    e. polycarpic fruit

20. The blackberry is composed of single-seeded, fleshy parts derived from a single flower. How is the blackberry fruit classified?
    a. simple fruit
    b. aggregate fruit
    c. multiple fruit
    d. vegetative fruit
    e. parthenocarpic fruit

21. Which DOES NOT describe bean seedling development?
    a. Its development is representative of dicots.
    b. The radicle is the first embryonic part to emerge from the seed.
    c. The radicle provides a source of food until the seedling begins photosynthesis.
    d. The hypocotyl is bent until the shoot breaks through the soil surface.
    e. The cotyledons drop off after the seedling becomes photosynthetic.

22. Which DOES NOT describe corn seedling development?
    a. The cotyledon absorbs nutrients from the endosperm.
    b. The epicotyl is enclosed by a protective sheath.
    c. Corn seedling development is representative of monocots.
    d. After emergence of the epicotyl, the cotyledons wither and fall off.
    e. The seedling becomes autotrophic after emergence of the epicotyl and development of foliage leaves.

Concept Map Construction

Construct a concept map for each group of terms. Be sure to include appropriate connector phrases. You may add other terms as necessary and use terms in the singular or plural form.

1. petal, incomplete flower, perfect flower, ovule, leaf
2. male gametophyte, megaspore, meiosis, aggregate fruit, monoecious
3. embryo sac, cotyledon, xylem, seed dormancy, parthenocarpy

Chapter 21

How Plants Grow and Develop

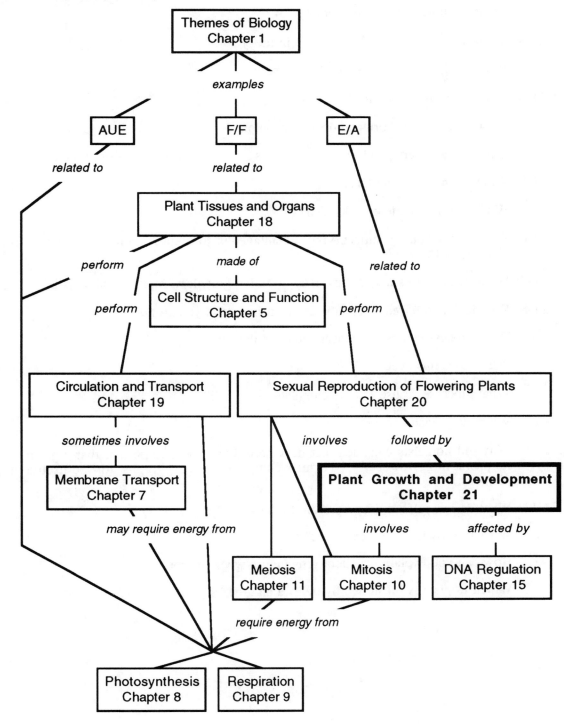

Go Figure!

Figure 21-1: Which aspect of pine growth not affected by the environment is also visible in Figure 21-1b?

Which hormone is responsible for this?

Figure 21-2: What do the colored dots represent in the shoot?

What is shown in the "exploded" box?

Figure 21-3: What do the colored dots represent in the shoot?

What is shown in the "exploded" boxes?

Figure 21-4: What does IAA stand for?

To which class of hormone does IAA belong?

Why do gardeners pinch the tips of stems to produce fuller growth in their plants?

Figure 21-5: What are nodes on a plant?

Where are internode cells located?

Figure 21-6: Why is thickening and curving advantageous to stems when they encounter an obstacle in the soil?

Figure 21-7: What three factors may be involved in timing growth with respect to season?

Figure 21-8: Why is P_{fr} referred to as "the active form" of phytochrome?

What process causes P_{fr} to accumulate in the cell?

When does this occur?

What processes cause P_r to accumulate in the cell?

When does each occur?

Figure 21-9: Why did botanists conclude that dark period was more important than light period in promoting flowering? (Hint: compare the bar graphs for poinsettia and hibiscus.)

Figure 21-11: What do the colored dots represent in the shoot?

What is shown in the "exploded" box?

How can you explain the shoot growing up when the root grows down?

Figure 21-12: Which tropism is illustrated by the pea and the Virginia creeper?

Matching Questions

Write the letter of the phrase that best matches the numbered term on the left. Use each only once.

_____ 1. gibberellin

_____ 2. ethylene gas

_____ 3. abscisic acid

_____ 4. apical dominance

_____ 5. senescence

_____ 6. photoperiod

_____ 7. vernalization

_____ 8. long-day plant

_____ 9. short-day plant

_____10. etiolation

_____11. P_r

_____12. P_{fr}

_____13. endogenous rhythm

_____14. circadian cycle

_____15. phototropism

_____16. negative gravitropism

_____17. positive gravitropism

_____18. thigmotropism

a. growth towards a light source

b. the active form of phytochrome

c. growth towards a gravitational force

d. flowering occurs when days are longer than some critical minimum length

e. this involves a 24-hour interval

f. mobilizes food resources in the seed

g. promotes etiolation when in high concentration

h. an internally controlled cycle

i. length of daytime in a 24-hour day

j. flowering occurs when day length is shorter than some critical maximum length

k. a response to gravity by stems

l. a growth response initiated by contact with another object

m. a phenomenon delayed by cytokinins

n. a phenomenon associated with indole acetic acid

o. promotes fruit ripening

p. inhibits cell division leading to plant dormancy

q. flowering only occurs after a cold period

r. process by which a plant becomes whitish due to a lack of chlorophyll

Multiple Choice Questions

1. Which is a *group of chemicals* known to act as plant hormones?
   a. indole acetic acid
   b. gibberellin
   c. ethylene gas
   d. sucrose
   e. phytochrome

2. Which is a *specific molecule* known to act as a plant hormone?
   a. auxin
   b. cytokinin
   c. abscisic acid
   d. P_{fr}
   e. potassium

3. Which is NOT an action of auxin?
   a. It stimulates axillary bud development.
   b. It stimulates elongation of cells.
   c. It is involved in the phototropic response.
   d. It promotes flowering.
   e. It stimulates mitosis in vascular cambium.

4. Which is NOT an action of gibberellin?
   a. It inhibits seed formation.
   b. It promotes pollen tube development.
   c. It makes fruit grow larger.
   d. It promotes dormancy.
   e. It inhibits seed formation.

5. Which is an action of ethylene gas?
   a. It is involved in the phototropic response.
   b. It promotes fruit ripening.
   c. It stimulates rapid cell division in stems.
   d. It stimulates rapid cell division in roots.
   e. It is active in wound healing.

6. Which is an action of abscisic acid?
   a. It promotes elongation of stem cells not receiving direct illumination.
   b. It promotes elongation of stem cells receiving direct illumination.
   c. It stimulates fruit ripening.
   d. It prolongs the life of leaves and flowers.
   e. It promotes dormancy.

7. Which DOES NOT describe auxin?
   a. One kind has been identified as indole acetic acid.
   b. It is classified as a plant hormone.
   c. It is responsible for apical dominance.
   d. Its activities can be performed by synthetic growth regulators.
   e. Large concentrations are required to produce physiological effects.

8. Which DOES NOT describe cytokinin?
   a. It was first discovered in coconut milk and yeast extract.
   b. It delays senescence.
   c. It stimulates rapid cell division in stems.
   d. It decreases transport of nutrients within the plant.
   e. It is involved in wound repair.

9. Which is used by plants to coordinate their development with environmental changes?
    a. temperature
    b. humidity
    c. insect pollinators
    d. wind
    e. spines

10. Which is used by plants to coordinate their development with environmental changes?
    a. fruit production
    b. waxy cuticle
    c. photoperiod
    d. photo opportunity
    e. root hair production

11. Which DOES NOT describe phytochrome?
    a. It is a plant pigment.
    b. It exists in two forms.
    c. P_r absorbs longer wavelength light than P_{fr}.
    d. P_r converts to P_{fr} when it absorbs light.
    e. P_{fr} accumulates during the daytime.

12. Which DOES NOT describe phytochrome?
    a. P_{fr} is the biologically active form.
    b. P_{fr} converts to P_r in dark periods.
    c. P_{fr} converts to P_r when long wavelength (far-red) light is absorbed.
    d. It is involved in the phototropic response.
    e. It is involved in measuring day length.

NOTE: *Answers for questions 13 and 14 are provided in the Appendix. Questions 15 and 16 constitute a similar sequence, but the answers are not given.*

13. The poinsettia is a short-day plant. Which describes poinsettias?
    a. They only flower if days are short and nights are long.
    b. They only flower if days are shorter than nights.
    c. They only flower if days are shorter than some critical maximum length.
    d. They only flower if days are longer than some critical minimum length.
    e. They flower regardless of day length.

14. The poinsettia is a short-day plant. Which describes poinsettias?
    a. They flower in February.
    b. They flower in spring.
    c. They flower in summer.
    d. They flower in autumn.
    e. They flower in response to cold temperatures.

15. Clover is a long-day plant. Which describes clover?
    a. It only flowers if days are long and nights are short.
    b. It only flowers if days are longer than nights.
    c. It only flowers if days are longer than some critical minimum length.
    d. It only flowers if days are shorter than some critical maximum length.
    e. It flowers regardless of day length.

16. Clover is a long-day plant. Which describes clover?
    a. It flowers in September.
    b. It flowers in November.
    c. It flowers in spring.
    d. It flowers in winter.
    e. It flowers in response to cold temperatures.

17. Which DOES NOT describe phototropism?
    a. Leaves demonstrate the phototropic response.
    b. Stems demonstrate the phototropic response.
    c. Production of an auxin concentration gradient is involved.
    d. Cells elongation occurs on the side of the plant away from the light source.
    e. Phytochrome is the pigment involved in detecting the light source.

18. Which DOES NOT describe gravitropism?
    a. Detection of the stimulus occurs by particles settling in the cell.
    b. Stems exhibit negative gravitropism.
    c. Roots exhibit positive gravitropism.
    d. The root cap produces a hormone that promotes cell elongation.
    e. Production of an auxin concentration gradient is involved.

Concept Map Construction

Construct a concept map for each group of terms. Be sure to include appropriate connector phrases.
You may add other terms as necessary and use terms in the singular or plural form.

1. auxin, abscisic acid, P_{fr}, photoperiodism, senescence
2. apical dominance, DNA, hormone, gibberellin, ovary
3. phytochrome, cotyledon, florigen, chloroplast, long-day plant

Chapter 22

An Introduction To Animal Form and Function

Section Concept Map

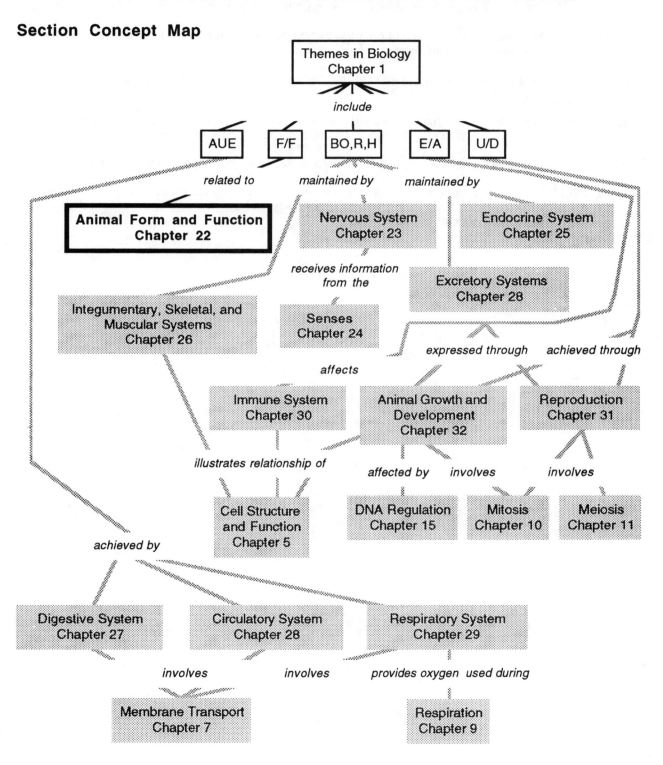

Go Figure!

Figure 22-1:   In what way are sweat glands and skeletal muscles antagonistic in the maintenance of human body temperature?

Which part of the human body is analogous to the home heater in temperature regulation?

Figure 22-2:   Why do you think the authors of your text have chosen to introduce muscle *and* nerve tissue in the same illustration?

Figure 22-8:   What is meant by the statement that the stapes can be *traced* to a bone in the skull of ancestral fishes?

Figure 22-10: According to the figure, does volume or surface area increase the fastest as you increase the size of a sphere?...decrease the fastest as you decrease the size of a sphere?

If a spherical cell was flattened into a pancake, how would that affect its surface area?

Figure 22-11: How do you think *total* oxygen consumed per minute by the mouse compares to that of the elephant? Don't confuse per gram body weight consumption with total oxygen consumed per unit of time.

Matching Questions

Write the letter of the phrase that best matches the numbered term on the left. Use each only once.

_____ 1. homeostasis

_____ 2. negative feedback mechanism

_____ 3. human body effectors

_____ 4. positive feedback mechanism

_____ 5. tissue

_____ 6. epithelial tissue

_____ 7. epidermis

_____ 8. connective tissue

_____ 9. collagen

_____ 10. cornea

_____ 11. blood vessel constriction

_____ 12. nerve tissue

_____ 13. organ

_____ 14. immune system

_____ 15. neuron

_____ 16. hormones

_____ 17. homologous

_____ 18. analogous

a. a change in one variable accelerates further change in that variable

b. epithelium forming outer layer of human skin

c. a sheet of cells lining body spaces

d. a change in one variable reduces further change in that variable

e. connective tissue protein

f. a function of muscle tissue

g. general name for any group of cells with a related structure and common function

h. maintenance of conditions within narrow limits

i. connective tissue covering the eye

j. includes muscles and glands

k. loosely organized cells within a nonliving matrix

l. chemicals produced by the endocrine system of animals

m. term used to describe similar structures derived from a common ancestral organ

n. includes very long cells along which an electrical signal is sent

o. discrete structure, such as the brain, composed of tissues

p. major cell type within the nervous system

q. protects an animal from foreign substances which have entered its body

r. ter used to describe similar structures found in two species which was not derived from a common ancestral organ

Multiple Choice Questions

1. The human body maintains temperature homeostasis by
   a. a positive feedback mechanism.
   b. a negative feedback mechanism.
   c. scaling effects.
   d. the immune system.
   e. temperature receptors which lower body temperature when it is too high.

2. The human body regulates blood clotting by means of
   a. a positive feedback mechanism.
   b. a negative feedback mechanism.
   c. scaling effects.
   d. an "effector" which lowers clotting rate after clotting has begun.
   e. a receptor which lowers clotting rate when it is too high.

3. The human body receives information about homeostatically maintained internal and external conditions through
   a. positive feedback mechanisms.
   b. negative feedback mechanisms.
   c. "sensors."
   d. body "integrators."
   e. "effectors."

4. To maintain homeostasis, the human body processes incoming information and sends signals to the body from
   a. positive feedback mechanisms.
   b. negative feedback mechanisms.
   c. "sensors."
   d. body "integrators."
   e. "effectors."

5. The lowest/simplest organizational level composed of a group of cells with related structure and function best describes
   a. a tissue.
   b. an organ.
   c. an organ system.
   d. an antibody.
   e. the scaling effect.

6. An organized aggregate of cell groups with different structures and functions working together to perform a particular task best describes
   a. a tissue.
   b. an organ.
   c. an organ system.
   d. an antibody.
   e. the scaling effect.

7. A sheet of tightly adhering cells best describes
   a. epithelial tissue.
   b. connective tissue.
   c. muscle tissue.
   d. nervous tissue.
   e. collagen.

8. A group of loosely organized cells surrounded by a nonliving extracellular matrix best describes
    a. epithelial tissue.
    b. connective tissue.
    c. muscle tissue.
    d. nervous tissue.
    e. collagen.

9. A group of cells containing elongated protein filaments which allow each cell to contract best describes
    a. epithelial tissue.
    b. connective tissue.
    c. muscle tissue.
    d. nervous tissue.
    e. collagen.

10. A group of cells capable of sending a "signal" along their membrane to other cells best describes
    a. epithelial tissue.
    b. connective tissue.
    c. muscle tissue.
    d. nervous tissue.
    e. collagen.

11. Functions of epithelial tissues include
    a. cell contraction to effect movement of the organism.
    b. formation of an intracellular matrix composed of protein and polysaccharide.
    c. protection of underlying tissues.
    d. communication between different areas of an animal's body.
    e. blood formation.

12. Functions of epithelial tissues include
    a. changing the shape of the human eye lens.
    b. secretion of materials into body spaces.
    c. movement of food through the digestive tract.
    d. formation of the lower layer of the skin (dermis).
    e. constriction of blood vessels.

13. Functions of muscle tissue include
    a. formation of the outer layer of the eyeball.
    b. the formation of an intracellular matrix composed of protein and polysaccharide.
    c. movement of food through the digestive tract.
    d. communication between different areas of an animal's body.
    e. the formation of blood.

14. Functions of muscle tissue include
    a. production of hormones.
    b. production of digestive enzymes.
    c. constriction of blood vessels.
    d. formation of collagen.
    e. transport of fat to adipose tissue.

15. Functions of a urinary system include
    a. absorption of macromolecules produced by digestion.
    b. ridding the body of metabolic waste products.
    c. maintenance of body shape.
    d. antibody production.
    e. oxygen absorption.

16.   Functions of a urinary system include
    a.   expulsion of carbon dioxide gas.
    b.   transport of oxygen gas.
    c.   storage of gametes until they are needed.
    d.   regulation of salt and water content.
    e.   providing rigid structures upon which muscles can exert a force.

17.   Functions of a skeletal system include
    a.   absorption of macromolecules produced by digestion.
    b.   ridding the body of metabolic waste products.
    c.   maintenance of body shape.
    d.   antibody production.
    e.   oxygen absorption.

18.   Functions of a skeletal system include
    a.   expulsion of carbon dioxide gas.
    b.   transport of oxygen gas.
    c.   regulation of sodium chloride (table salt) and water content.
    d.   providing rigid structures upon which muscles can exert a force.
    e.   production of the body's hormones.

19.   Which animal has the greatest surface area to volume ratio?
    a.   house fly
    b.   human
    c.   dog
    d.   cow
    e.   pigeon

20.   Based on size alone, which animal will have the least trouble maintaining a constant body temperature on a cold winter morning?
    a.   house fly
    b.   human
    c.   dog
    d.   cow
    e.   pigeon

Concept Map Construction

Construct a concept map for each group of terms. Be sure to include appropriate connector phrases. You may add other terms as necessary and use terms in the plural or singular form.

    1.   homeostasis, negative feedback mechanism, hormone, Golgi body, collagen
    2.   ribosome, contractile protein, tissue, epidermis, organ
    3.   plasma membrane, surface area to volume ratio, digestive system, respiratory system, homeostasis

Chapter 23

Coordinating The Organism:
The Role of the Nervous System

Section Concept Map

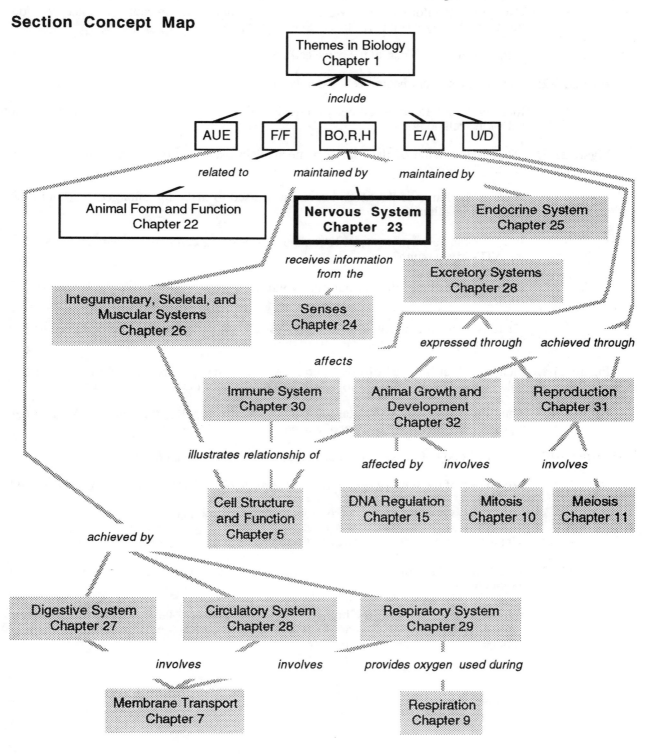

Go Figure!

Figure 23-1:   Assume a neural impulse begins at a dendrite and travels to a target cell. Name each neuron component over which it travels in reaching its goal.

What is found at the end of each axon collateral fiber?

Using just the information provided in this diagram and the preceding chapter, what do you think the target cell (muscle) response will be?

Is the myelin sheath part of the nerve cell?

Figure 23-2:   Identify the sensory neuron cell body. Is the axon to the left or right of the cell body?

Identify the interneuron cell body on the interneuron to the right. Is the axon above or below the cell body?

Figure 23-3:   What is the name of the transport mechanism used to move sodium ions out of a "resting" neuron?

What is the name of the neuron component used to make the enlarged section in the middle of this figure?

Why are there always plenty of K^+ on the outside of the cell to be exchanged for Na^+ by the ATP-driven pump?

What is the name of the mechanism by which K^+ move from inside the neuron to the outside?

There is a higher concentration of negatively charged ions inside a neuron than outside. Why don't protein anions move toward the outside?

Figure 23-4:   Please look at part "a" of this figure. In what ways does repolarization differ from polarization? Draw an arrow to indicate the direction of impulse propagation.

Figure 23-5:   This illustration shows skeletal muscle cells as the target of this neuron. What other types of cells could be the target of a neuron?

Figure 23-10:   Is the cell body of the sensory neuron found within the spinal cord?

Which part of the sensory neuron acts as the receptor? Which part enters the spinal cord?

Is the cell body of the motor neuron found within the spinal cord?

Which part of the motor neuron carries an impulse to the muscle?

Figure 23-13:   Which general region of the human body is controlled by the largest area of the brain's motor cortex?

Which general region of the human body has a relatively large area of motor cortex controlling it, but a small area of somatic sensory cortex used in receiving incoming impulses?

Figure 23-16:   Why isn't there a ventral root ganglion?

                     Cerebrospinal fluid is found between what two anatomical structures?

                     What bony structures surround and protect the spinal cord? Given this association, what might cause a "pinched" spinal nerve?

Figure 23-17:   During times of stress, what advantage does an organism gain by sympathetic stimulation reducing salivary gland secretions and inhibiting stomach and intestine muscular contractions?...by increasing the release of glucose from the liver?

Figure 23-18:   List all the components of the central nervous system shown in this illustration.

                     List all the components of the peripheral nervous system shown in this figure.

Matching Questions

Write the letter of the phrase that best matches the numbered term on the left. Use each only once.

_____ 1. neuron

_____ 2. cell body

_____ 3. dendrite

_____ 4. nervous impulse

_____ 5. target cell

_____ 6. Schwann cell

_____ 7. collateral fibers

_____ 8. sensory neuron

_____ 9. node of Ranvier

_____10. ionic gradient

_____11. anion

_____12. leak channel

_____13. polarized

_____14. depolarization

_____15. repolarization

_____16. saltatory conduction

_____17. acetylcholine

_____18. Ca^{++}

_____19. hyperpolarize

_____20. summate

a. narrow cytoplasmic processes which conduct neural impulses toward a nerve cell's cell body

b. forms myelin sheath

c. uninsulated gap of a myelinated neuron

d. a decrease in polarization

e. a neurotransmitter

f. negatively charged atom or molecule

g. to add together

h. a cell that responds to a neural impulse may be part of the nervous, muscular, or endocrine system

i. type of nerve cell that carries impulses towards the CNS

j. a nerve cell

k. intramembrane passage that is always open

l. stimulates exocytosis

m. reestablish the original polarized state

n. increasing the magnitude of difference between two charges

o. small branches from an axon

p. contains the nucleus of a nerve cell

q. possessing opposite charges

r. modification of neural impulse resulting from leaps between nodes of Ranvier

s. a difference in charged particle concentration between two regions

t. wave of depolarization traveling along a nerve cell

Multiple Choice Questions

1. The human nervous system
   a. regulates heart rate.
   b. removes metabolic wastes.
   c. consists of modified epithelial cells.
   d. contains a single type of cell.
   e. forms the lining of the body cavities.

2. The human nervous system
   a. forms the outer covering of the eye.
   b. helps coordinate homeostasis.
   c. forms by cephalization.
   d. is composed of molecular subunits called nodes of Ranvier.
   e. is analogous to the nerve net of an anemone.

3. Effectors of the human body include
   a. dendrites.
   b. Schwann cells.
   c. nodes of Ranvier.
   d. muscle cells.
   e. connective tissue cells.

4. Effectors of the human body include
   a. cerebrospinal fluid.
   b. neurons.
   c. neuroglial cells.
   d. anions.
   e. epithelial cells.

5. Neurons carrying impulses towards the central nervous system best describes
   a. excitatory neurons.
   b. inhibitory neurons.
   c. sensory neurons.
   d. motor neurons.
   e. interneurons.

6. Neurons carrying impulses away from the central nervous system best describes
   a. excitatory neurons.
   b. inhibitory neurons.
   c. sensory neurons.
   d. motor neurons.
   e. interneurons.

7. What is a primary function of Schwann cells?
   a. They surround and protect a neuron's dendrites.
   b. They surround and protect a neuron's synaptic cleft.
   c. They act to repolarize a neuron's membrane.
   d. They act to insulate the charge of a neuron from that of its neighbors.
   e. They allow the neuron's polarization to be negated by surrounding neurons.

8. What is a primary function of Schwann cells?
   a. They surround and protect a neuron's cell body.
   b. They surround and protect a neuron's synaptic vesicles.
   c. They speed the rate of nerve impulse conduction.
   d. They slow the rate of nerve impulse conduction.
   e. They depolarize the neuron's membrane.

9. Which is true of a neuron at "rest?"
   a. It has a lower concentration of potassium ions inside than out.
   b. It has a higher concentration of potassium ions inside than out.
   c. It has a higher concentration of sodium ions inside than out.
   d. It has a higher concentration of chloride ions inside than out.
   e. It has a lower concentration of protein anions inside than out.

10. Which is true of a neuron at "rest?"
    a. It has sodium ion leak channels.
    b. It has open sodium ion gated channels.
    c. It has closed sodium ion gated channels.
    d. It has open potassium ion gated channels
    e. It has a membrane potential of approximately +70 millivolts.

11. Which is true of a neuron during depolarization?
    a. Potassium ion gated channels open.
    b. Potassium ion gated channels close.
    c. Sodium ion leak channels open.
    d. Sodium ion leak channels close.
    e. Sodium ion gated channels open.

12. Which is true of a neuron during depolarization?
    a. The charge across the membrane increases.
    b. The charge across the membrane decreases.
    c. The charge across the membrane stays the same.
    d. Sodium ions leave the interior.
    e. Potassium ions stop moving across leak channels.

13. Synaptic knobs associated with a synaptic cleft function to
    a. receive impulses from neighboring neurons.
    b. release neurotransmitter substances into the synaptic cleft.
    c. repolarize the dendrite to which it is attached.
    d. repolarize the axon to which it is attached.
    e. stimulate hyperpolarization in the dendrite to which it is attached.

14. Synaptic knobs associated with a synaptic cleft function to
    a. store neurotransmitter substances in membrane vesicles.
    b. stimulate hyperpolarization in the axon to which it is attached.
    c. insulate the cell body from neighboring neurons.
    d. depolarize the dendrite to which it is attached.
    e. depolarize the axon to which it is attached.

15. Which is a neurotransmitter?
    a. anion
    b. acetylcholine
    c. Ca^{++}
    d. Na^{+}
    e. K^{+}

16. Which is a neurotransmitter?
    a. cation
    b. Cl^{-}
    c. norepinephrine
    d. gray matter
    e. myelin

17. Summation may result from
    a. many simultaneous inhibitory signals from adjacent neurons.
    b. continuous, repeated inhibitory signals from one adjacent neuron.
    c. many simultaneous excitatory signals from adjacent neurons.
    d. continuous, repeated excitatory signals from one adjacent neuroglial cell.
    e. the loss of the myelin sheath.

18. Summation may result from
    a. many simultaneous inhibitory signals from adjacent Schwann cells.
    b. continuous, repeated excitatory signals from one adjacent neuron.
    c. continuous, repeated inhibitory signals from one adjacent neuroglial cell.
    d. the addition of the myelin sheath.
    e. many simultaneous excitatory signals from neuroglial cells.

19. Which is a component of every reflex?
    a. interneuron
    b. neuroglial cell
    c. brain
    d. motor cortex
    e. sensory neuron

20. Which is a component of every reflex?
    a. sensory cortex
    b. cerebral cortex
    c. spinal cord
    d. motor neuron
    e. ventral root ganglion

21. Interconnected chambers within the brain are called
    a. motor cortices
    b. pons
    c. ventricles
    d. cerebrospinal fluid
    e. corpus callosum

22. What structure connects the two cerebral hemispheres?
    a. motor cortices
    b. pons
    c. ventricles
    d. cerebrospinal fluid
    e. corpus callosum

Concept Map Construction

Construct a concept map for each group of terms. Be sure to include appropriate connector phrases. You may add other terms as necessary and use terms in the plural or singular form.

1. plasma membrane, myelin sheath, neuroglial cell, node of Ranvier, motor neuron
2. active transport, polarization, potassium ion, muscle tissue, membrane channel
3. somatic nervous system, spinal nerve, sensory receptor, parasympathetic nervous system, neurotransmitter

Chapter 24

Sensory Perception:
Gathering Information About The Environment

Section Concept Map

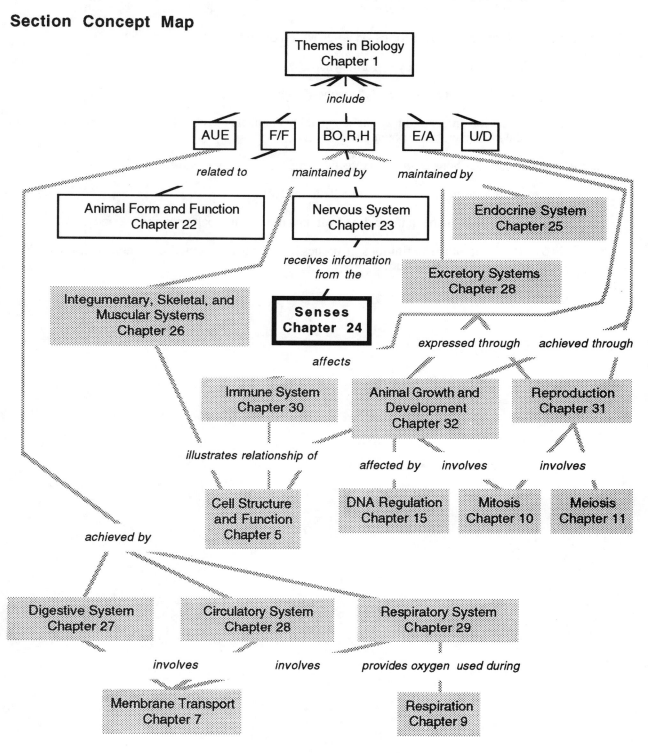

Go Figure!

Figure 24-3:   Is this cortex involved in receiving and interpreting sensory information at a conscious or unconscious level?

                Which general part of the body has the largest amount of sensory cortex involved in receiving and interpreting nervous impulses? What do you think this means about the number of sensory receptors in that body region compared with other regions?

Figure 24-4:   Which retinal layer forms synapses with rods and cones?

                How many retinal layers form synapses?

                Which part of a rod actually "receives" light?

                What ion enters a rod when the gated ion channels open?...What are the consequences to the rod's membrane potential?

Figure 24-5:   Which ear bone is attached directly to the tympanic membrane?

                Which ear bone is attached directly to the stirrup?

                What might happen if the round window did not dissipate incoming sounds?

Figure 24-7:   Based upon the definitions of tissues and organs presented in Chapter 22, how would you classify a taste bud?

Matching Questions

Write the letter of the phrase that best matches the numbered term on the left. Use each only once.

_____ 1. stimulus

_____ 2. sensory receptor

_____ 3. mechanoreceptor

_____ 4. Pacinian corpuscle

_____ 5. rod cell

_____ 6. sclera

_____ 7. ciliary muscles

_____ 8. accommodation

_____ 9. retina

_____10. retinal

_____11. dark adaptation of the eye

_____12. tapetum

_____13. tympanic membrane

_____14. stirrup

_____15. basilar membrane

_____16. auditory nerve

_____17. sweet

_____18. taste bud

_____19. fovea

_____20. outer ear

a. nonspecific category of cells capable of initiating a nervous impulse in response to a change in the environment

b. an example of a photoreceptor cell

c. an example of connective tissue

d. includes rod pigment activation

e. all vertebrate photoreceptor cells are found here

f. an example of a mechanoreceptor

g. location of cochlea's hair cells

h. includes the pinna

i. contains neurons running from the cochlea to the brain

j. the act of changing the shape of the eye's lens

k. light reflective layer behind the retina

l. any change in the environment

m. one of the primary tastes

n. bone attached to the cochlea

o. portion of retina containing the greatest concentration of cones

p. located at the inner end of the ear canal

q. one location of chemoreceptors

r. a cell that may respond specifically to touch

s. nonprotein portion of light sensitive pigment in the human eye

t. change the shape of the eye's lens

Multiple Choice Questions

1. Which is considered a sensory receptor?
   a. sclera
   b. choroid
   c. Pacinian corpuscle
   d. basilar membrane
   e. somatic sensory cortex

2. Which is considered a sensory receptor?
   a. cornea
   b. pupil
   c. cone cells
   d. rhodopsin
   e. tapetum

3. Mechanoreceptors respond to
   a. changes in temperature.
   b. light.
   c. chemicals that are released from damaged tissue.
   d. changes in pressure.
   e. chemicals in the air.

4. Some chemoreceptors respond to
   a. changes in temperature.
   b. light.
   c. changes in temperature.
   d. changes in pressure.
   e. chemicals in the air.

5. The outermost layer of the eye is the
   a. choroid.
   b. retina.
   c. optic nerve.
   d. sclera and cornea.
   e. iris and lens.

6. The innermost layer of the eye is the
   a. choroid.
   b. retina.
   c. optic nerve.
   d. sclera and cornea.
   e. iris and lens.

7. How does the human eye focus an image on the retina?
   a. As light passes through the sclera, light rays are focused by a change in the shape of the cornea.
   b. As light passes through the iris, light rays are focused by a change in the size of the pupil.
   c. As light passes through the lens, the rays are focused by a change in the shape of the lens.
   d. As light passes through the vitreous humor, the rays are focused on the retina.
   e. As light strikes the bipolar layer it is reflected back to the rod and cone layer where it is focused.

8. How is the shape of the human lens changed?
   a. The ganglion cell layer contracts and changes the lens' shape.
   b. Ciliary muscles contract and change the lens' shape.
   c. Ligaments attached to the lens contract and change its shape.
   d. The sclera contracts and changes the lens' shape.
   e. The muscles of the iris contract and change the lens' shape.

9.   Which type of eye cell is responsible for color perception?
   a.   cone
   b.   ciliary muscle cell
   c.   cells of the bipolar layer
   d.   cells of the ganglion layer
   e.   rod

10.   Which type of eye cell is responsible for black and white vision?
   a.   cone
   b.   ciliary muscle cell
   c.   cells of the bipolar layer
   d.   cells of the ganglion layer
   e.   rod

11.   All human visual pigments contain
   a.   rhodopsin.
   b.   a protein portion.
   c.   neurotransmitters.
   d.   acetylcholine.
   e.   vitamin A.

12.   All human visual pigments contain
   a.   vitreous humor.
   b.   norepinephrine.
   c.   tapetum.
   d.   retinal.
   e.   chemoreceptors.

13.   In the human eye, which occurs during dark adaptation?
   a.   ciliary muscles contract
   b.   inactivation of rhodopsin
   c.   closure of the iris
   d.   activation of cone pigments
   e.   activation of rod pigments

14.   In the human eye, which occurs during dark adaptation?
   a.   ciliary muscles relax
   b.   inactivation of vitreous humor
   c.   dilation of the iris
   d.   inactivation of cone pigments
   e.   inactivation of rod pigments

15.   What is the primary type of sensory receptor in the human ear?
   a.   rods
   b.   Meissner's corpuscles
   c.   hair cells
   d.   vestibular apparatus
   e.   thermoreceptor

16.   What are the two primary sensory systems in the human ear?
   a.   olfaction and maintenance of posture
   b.   thermoreception and maintenance of posture
   c.   awareness of tissue damage and thermoreception
   d.   hearing and maintenance of posture
   e.   hearing and maintenance of balance

17. What comprises the human middle ear?
    a. a chamber and three ear bones
    b. a chamber and the vestibular apparatus
    c. the cochlea and the vestibular apparatus
    d. the pinna and a channel leading to the tympanic membrane
    e. the vestibular apparatus and the auditory nerve

18. What comprises the inner ear in humans?
    a. a chamber and three ear bones
    b. a chamber and the vestibular apparatus
    c. the cochlea and the vestibular apparatus
    d. the pinna and a channel leading to the tympanic membrane
    e. the vestibular apparatus and the auditory nerve

Concept Map Construction

Construct a concept map for each group of terms. Be sure to include appropriate connector phrases. You may add other terms as necessary and use terms in the plural or singular form.

1. mechanoreceptor, somatic sensory cortex, sensory neuron, somatic nervous system, retina
2. hypothalamus, thermoreceptor, autonomic nervous system, hair cell, vestibular apparatus
3. dark adaptation, visual pigment, rhodopsin, visual cortex, blind spot of the eye

Chapter 25

Coordinating The Organism:
The Role Of The Endocrine System

Global Concept Map

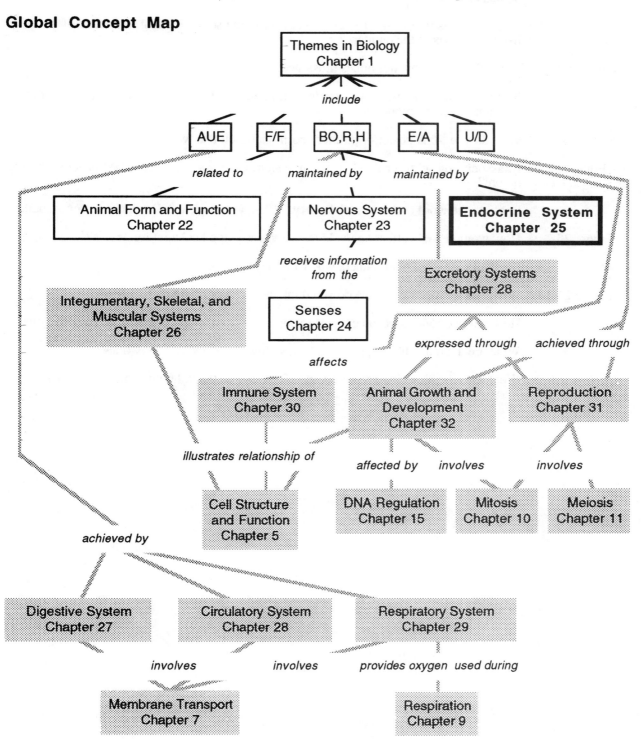

Go Figure!

Figure 25-5:   In what way does the anterior pituitary capillary arrangement differ from capillary structure in the posterior pituitary? Base your answer only upon the figure.

What type of cell is present in the anterior pituitary and missing from the posterior pituitary?

From the illustration, what is the first event in the series leading to release of hormone from both parts of the pituitary?

Figure 25-6:   Name the four tropic hormones illustrated.

Name the anterior pituitary hormone stimulating release of thyroxine...progesterone... stimulating sperm production.

Why do you think posterior pituitary hormones aren't shown?

Figure 25-9:   In association with what other organ system have you seen the word norepinephrine before?

Look up the word cortex in a dictionary. What does it mean? Do the same for medulla.

Figure 25-11:   How would PTH's affect on the kidneys increase blood calcium ion levels?

If calcium ions are not retained, where do they end up?

Figure 25-14:   What type of macromolecule do you think the receptor might be?

How might cAMP affect an enzyme to convert it from an inactive state to an active one?

What are the molecules represented by a yellow circle and two gray lines?

What does the blue circle represent in "b?" What do the light blue parts represent?

Matching Questions

Write the letter of the phrase that best matches the numbered term on the left. Use each only once.

_____ 1.   adrenocorticotropic hormone

_____ 2.   cortisol

_____ 3.   insulin

_____ 4.   melatonin

_____ 5.   neurosecretory cells

_____ 6.   posterior pituitary gland

_____ 7.   releasing factor

_____ 8.   antidiuretic hormone

_____ 9.   somatotropin

_____10.   acromegaly

_____11.   mineralcorticoids

_____12.   adrenal gland

_____13.   Addison's disease

_____14.   adrenalin

_____15.   thyroid gland

_____16.   triiodothyronine

_____17.   hypothyroidism

_____18.   Type I diabetes

_____19.   Type II diabetes

_____20.   prostaglandin

a.   stimulative product of hypothalamic neurosecretory cells--is not stored in the posterior pituitary

b.   glucocorticoid hormone of the adrenal cortex

c.   produced by cells originating in the hypothalamus and reduces water loss from the kidneys

d.   results from too much growth hormone

e.   modified fatty acids

f.   an amino acid containing iodine

g.   results from insulin receptor abnormalities

h.   aldosterone is one of these

i.   hormone produced by the pineal gland

j.   stores hormones produced by cells originating in the hypothalamus

k.   also known as epinephrine

l.   begins during childhood

m.   look like neurons, function like endocrine cells

n.   in the next near the larynx

o.   characterized by extreme weakness and weight loss

p.   growth hormone

q.   one found on top of each kidney

r.   abnormally low levels of thyroid hormone

s.   directly regulates secretions of the adrenal cortex

t.   hormone produced in the pancreas

Multiple Choice Questions

1. In what way are the nervous and endocrine systems similar?
    a. Both use specific cellular paths to communicate with target cells.
    b. Both use chemical messengers produced by one cell and transported to a distant target cell.
    c. Both affect specific target cells.
    d. Both evoke rapid responses.
    e. Both exclusively maintain homeostasis.

2. In what way are the nervous and endocrine systems similar?
    a. Both consist of organs interconnected only by the circulatory system.
    b. Both produce chemical messengers with specific target cells immediately neighboring the cell that produced the messenger.
    c. Both maintain a -70 millivolt charge across the cells comprising the system.
    d. Both use chemical messengers.
    e. Both evoke long term changes in target cell metabolism.

3. In addition to the production of hormones, the pancreas also produces
    a. neurotransmitters.
    b. digestive enzymes.
    c. glucocorticoids.
    d. vitreous humor.
    e. Meissner's corpuscles.

4. Bayliss and Starling joined the circulatory system of two dogs and fed one of them. The result was the pancreas of both dogs secreted materials into the intestine. They concluded that
    a. food in the stomach was not necessary to produce pancreatic secretions.
    b. a chemical message was carried in the blood which signaled the pancreas to release its secretions.
    c. when a dog sees food, its pancreas releases secretions into the intestine.
    d. the pancreas produces secretin.
    e. neurons release materials into the blood.

5. Many known hormones are
    a. carbohydrates.
    b. phospholipids.
    c. nucleic acids.
    d. derivatives of amino acids.
    e. derivatives of carbohydrates.

6. Many known hormones are
    a. lipoproteins.
    b. steroids.
    c. derivatives of RNA.
    d. triglycerides.
    e. glycoproteins.

7. The hypothalamus
    a. acts as a link between the nervous and endocrine systems.
    b. is an important source of adrenocorticotropic hormone.
    c. responds to releasing factors produced by the pituitary gland.
    d. contains neurosecretory cells originating in the pituitary.
    e. receives blood through a system of capillaries after the blood has passed through the pituitary.

8. The posterior pituitary gland
   a. acts as a link between the nervous and endocrine systems.
   b. is an important source of adrenocorticotropic hormone.
   c. responds to releasing factors produced by the hypothalamus.
   d. contains neurosecretory cells originating in the hypothalamus.
   e. receives blood through a system of capillaries after the blood has passed through the anterior pituitary.

9. The hypothalamus monitors the condition of the body by means of
   a. sensory neurons of the somatic nervous system.
   b. blood flowing through it.
   c. motor neurons of the parasympathetic nervous system.
   d. motor neurons of the sympathetic nervous system.
   e. neurosecretory cells.

10. The hypothalamus monitors the condition of the body by means of
    a. sensory neurons of the spinal cord.
    b. blood flowing through the pituitary gland.
    c. sensory neurons.
    d. motor neurons of the central nervous system.
    e. secretory cells of the parathyroids.

11. Oxytocin
    a. stimulates milk release from the breasts.
    b. promotes milk production by the breasts.
    c. promotes increased muscle mass.
    d. stimulates release of pituitary hormones.
    e. reduces metabolic rate.

12. Growth hormone
    a. stimulates the release of milk from the breasts.
    b. promotes the production of milk by the breasts.
    c. promotes increased muscle mass.
    d. stimulates release of pituitary hormones.
    e. reduces metabolic rate.

13. The adrenal glands release glucocorticoids in response to increased blood levels of
    a. thyroid-stimulating hormone.
    b. gonadotrophic hormones.
    c. cortisol.
    d. adrenocorticotropic hormone.
    e. renin.

14. Sex hormone and gamete production are promoted by
    a. thyroid-stimulating hormone.
    b. gonadotrophic hormones.
    c. cortisol.
    d. adrenocorticotropic hormone.
    e. renin.

15. An effect of cortisol is
    a. increased loss of water through the kidneys.
    b. increased retention of sodium in the kidneys.
    c. suppression of the body's response to injury.
    d. increased mobility of stomach contents.
    e. sexual maturation.

16. An effect of aldosterone is
    a. increased loss of water through the kidneys.
    b. increased retention of sodium in the kidneys.
    c. suppression of the body's response to injury.
    d. increased mobility of stomach contents.
    e. sexual maturation.

17. The adrenal medulla secretes epinephrine in response to
    a. increased blood levels of adrenocorticotropic hormone.
    b. stimulation by sympathetic neurons.
    c. stimulation by parasympathetic nerve fibers.
    d. increased blood levels of aldosterone.
    e. decreased blood levels of aldosterone.

18. Immediate effects of epinephrine secretion by the adrenal medulla include
    a. decreased heart rate.
    b. decreased metabolic rate.
    c. increased blood flow to the skin.
    d. increased contraction of intestinal muscles.
    e. increased number of cells in the blood.

19. Thyroxine and triiodothyronine are
    a. hormones of the parathyroid glands.
    b. amino acids that contain iodine.
    c. proteins that contain iodine.
    d. proteins that contain potassium.
    e. hormones of the adrenal medulla.

20. Extremely low levels of thyroxine and triiodothyronine during childhood result in
    a. acromegaly.
    b. Addison's disease.
    c. cretinism.
    d. Grave's disease.
    e. exophthalmia.

21. Insulin's primary effect is to
    a. increase heart rate.
    b. regulate sugar retention by the kidneys.
    c. regulate conversion of glycogen to glucose.
    d. regulate conversion of glucose to glycogen.
    e. stimulate cellular uptake of blood glucose.

22. Type II diabetes directly results from
    a. a failure of the pancreas to produce insulin.
    b. over production of insulin.
    c. insulin receptor abnormalities.
    d. a failure of the hypothalamus to monitor blood iodine levels.
    e. underproduction of prolactin.

23. If cAMP is considered a "second messenger," what is the "first messenger?"
    a. adenylate cyclase
    b. a hormone
    c. G protein
    d. ATP
    e. NADPH

24.   What converts ATP to cAMP in a "second messenger" mechanism of hormone action?
       a.   adenylate cyclase
       b.   a hormone
       c.   G protein
       d.   ATP
       e.   NADPH

Concept Map Construction

Construct a concept map for each group of terms. Be sure to include appropriate connector phrases. You may add other terms as necessary and use terms in the plural or singular form.

   1.   connective tissue, parathyroid gland, calcitonin, negative feedback, synaptic vesicle
   2.   neuron, neurosecretory cell, neurotransmitter, hormone, estrogen
   3.   releasing factor, antidiuretic hormone, acromegaly, suppression of the immune system, protein

Chapter 26

Protection, Support, And Movement:
The Integumentary, Skeletal, And Muscular Systems

Section Concept Map

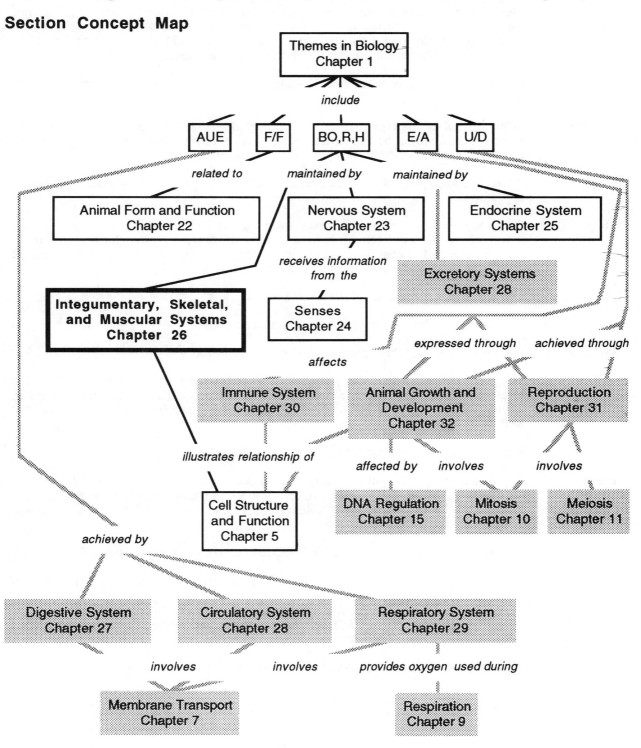

Go Figure!

Figure 26-1:   Which layer corresponds to what is commonly called skin?

In which skin layer do sensory receptor nerve endings appear? Do you think these are part of the somatic or visceral sensory system?

Many muscles cause movement because they are attached at one end to something rigid (or something held firmly in place) and at the other to something that can move. Contraction results in movement at the mobile end. Since the muscle anchored to a hair shaft is not attached to something immobile, what is the effect of its contraction?

By what mechanism do nutrients and wastes move between cells of the dermis and living cells of the lower epidermis?

Figure 26-4:   Which part of the appendage in "b" must be rigid for contraction to produce movement? If the upper and lower part of the appendage were not connected with a flexible piece of exoskeleton, what would happen when the flexor muscle contracted?

What functional purpose does the presence of flexor <u>and</u> extensor muscles serve in this appendage?

Figure 26-6:   Osteocytes remain a living part of bone even after bone elongation has stopped. But the living cells of hair or fingernails become filled with keratin and die to become part of the hair or fingernail. What functional purpose does the maintenance of living osteocytes serve in human bone?...How is this function served in hair and fingernails?

Figure 26-7:   Identify at least three bones of the axial skeleton.

Identify at least three bones of the appendicular skeleton.

Figure 26-10: Use Figure 26-7 and this figure to answer these questions.

To which bones are the biceps attached? Which bone remains immobile when the forearm is flexed? Which is moved? The bone which is held immobile during this motion can be moved at other times. What keeps it immobile during flexion of the forearm?

Answer the same questions for extension of the forearm.

Figure 26-11: Strenuous, repeated, regular exercise over a period of months causes muscle cells to enlarge (hypertrophy), particularly in the presence of the steroid hormone testosterone. Which muscles might enlarge if you vigorously and regularly did push-ups? If you worked out on a stair-climber?

Which is the major shoulder muscle? Which is the major human calf muscle?

Figure 26-12: Describe a physical reason that I bands and H zones stain lighter than the outer edges of the A band.

Why do you suppose muscle fiber nuclei are located along the outer edge of the cytoplasm rather than deep within it?

What purpose is served by having more than one nucleus in each muscle fiber?

Figure 26-13: What happens to the distance between Z lines during contraction?...width of the A band's darkly staining portions?

Which filament type is most abundant in a single myofibril?

Figure 26-15: What is the tubular projection of the plasma membrane at the bottom-right of this figure?

Figure 26-17: In what visible ways do cardiac muscle fibers differ from the skeletal muscle fibers shown in Figure 26-12?

Figure 26-18: The wheelbarrow handle is the lever. How does the length of the lever affect the force which must be applied? If you lifted from a point on the handle closer to the load, would you have to apply more or less force to lift the same load?

Matching Questions

Write the letter of the phrase that best matches the numbered term on the left. Use each only once.

_____ 1.   hair shaft

_____ 2.   epidermis

_____ 3.   dermis

_____ 4.   adipose tissue

_____ 5.   keratin

_____ 6.   cuticle

_____ 7.   sebum

_____ 8.   dermal bone

_____ 9.   feather

_____10.   hydrostatic pressure

_____11.   flexor muscle

_____12.   extensor muscle

_____13.   yellow marrow

_____14.   Haversian canals

_____15.   crushing forces

_____16.   smooth muscle

_____17.   antagonistic muscle pair

_____18.   sarcomere

_____19.   I band

a.   living epithelial layer of skin

b.   specific name applied to narrow, sensitive flap of tissue at the back rim of a fingernail

c.   avian structure originating in the epidermis

d.   located in the hollow core of long bones

e.   causes a limb straightened at the joint to bend

f.   area where actin filaments are separated from myosin filaments

g.   proteinaceous mammalian structure derived from epithelial tissue embedded in the dermis

h.   one contractile unit of a myofibril

i.   nonfatty connective tissue layer of skin

j.   primitive connective tissue arising from the dermis which contains calcium and other salts

k.   causes a limb bent at the joint to straighten

l.   primary component of hair and fingernails

m.   controlled primarily by the motor portion of the autonomic nervous system

n.   lipid mixture which coats hair and skin

o.   found in the subcutaneous layer of the integumentary system

p.   allow a limb to be moved in two directions

q.   pressure caused by weight placed upon an object

r.   channels through compact bone

s.   water pressure

Multiple Choice Questions

1. Which is a layer of human skin?
    a. the layer of loose connective tissue below the dermis
    b. the layer of fat below the dermis
    c. the cutaneous layer
    d. the red marrow
    e. the sarcomere

2. Which is a layer of the human integument?
    a. the subcutaneous layer
    b. the sarcoplasmic reticulum
    c. the lines of compression
    d. the yellow marrow
    e. the H zone

3. Which organs are found in the human integument?
    a. blood vessels, muscles, and the basal portions of exocrine glands
    b. blood vessels, muscles, and the basal portions of endocrine glands
    c. ganglia, muscles, and the basal portions of endocrine glands
    d. ganglia, muscles, and the basal portions of hair follicles
    e. muscles and the epidermal portion of endocrine glands

4. Which tissues are found in human skin?
    a. epithelial and connective
    b. epithelial and muscle
    c. epithelial, connective, muscle, and nerve
    d. connective, muscle, and nerve
    e. muscle and nerve

5. What was a major disadvantage for fish possessing armored shields of dermal bone?
    a. They had difficulty maintaining water homeostasis, since dermal bone interfered with normal urinary system function.
    b. They had reduced mobility and buoyancy.
    c. They were subject to injury of layers above the shields, since bone did not protect the overlying subcutaneous layer.
    d. They required large amounts of hard to get, protein-rich food, since keratin was deposited in the armored shields.
    e. The weight of the plates made it difficult for the fish to close their jaws while feeding.

6. What is a major difference between the integument of reptiles and amphibians?
    a. Amphibians possess a thin gas-permeable integument, while reptiles do not.
    b. Amphibians possess a thick gas-impermeable integument to prevent drying, while reptiles do not.
    c. Gas exchange does not occur across the integument of amphibians, while it does across reptilian integument.
    d. The integument of amphibians is dry, while that of reptiles is not.
    e. The integument of amphibians has large scales, while that of reptiles does not.

7. Which layer of an arthropod's integument forms an exoskeleton?
    a. cuticle
    b. epidermis
    c. dermis
    d. subcutaneous layer
    e. connective tissue

8. Which is a non-living, primarily protein and chitin layer in arthropods?
   a. cuticle
   b. epidermis
   c. dermis
   d. subcutaneous layer
   e. connective tissue

9. What type of tissue is bone?
   a. epithelial
   b. connective
   c. muscle
   d. nerve
   e. epithelial and connective

10. What are the primary components of bone's extracellular matrix?
    a. keratin and calcium phosphate
    b. chitin and sodium phosphate
    c. collagen and calcium phosphate
    d. keratin and sodium phosphate
    e. collagen and sodium phosphate

11. The protein portion of bone
    a. gives it flexibility and the strength to resist tension forces.
    b. gives it rigidity and the strength to resist compressional forces.
    c. provides a flexible outer covering.
    d. fills the cavity of long bones such as the femur.
    e. provides a path along which materials may diffuse into osteocytes.

12. The salt portion of bone
    a. gives it flexibility and the strength to resist tension forces.
    b. gives it rigidity and the strength to resist compressional forces.
    c. provides a flexible outer covering.
    d. fills the cavity of long bones such as the femur.
    e. provides a path along which materials may diffuse into osteocytes.

13. Which is a bone cell?
    a. chondrocyte
    b. osteocyte
    c. epidermal cell
    d. neuron
    e. collagen

14. Which is a cartilage cell?
    a. chondrocyte
    b. osteocyte
    c. epidermal cell
    d. neuron
    e. collagen

15.   Which is a major difference between bone and cartilage?
   a.   The matrix of bone is largely collagen and an amorphous composite of other proteins and polysaccharides. The matrix of cartilage is primarily collagen with calcium carbonate deposited between the fibers.
   b.   The matrix of bone is primarily chitin and calcium carbonate. The matrix of cartilage is largely collagen with calcium carbonate deposited between the fibers.
   c.   The matrix of bone is largely collagen and calcium phosphate. The matrix of cartilage is collagen with calcium carbonate deposited between the fibers.
   d.   The matrix of bone is largely collagen and calcium phosphate. The matrix of cartilage is collagen with an amorphous composite of other proteins and polysaccharides.
   e.   The matrix of bone is largely collagen and calcium phosphate. The matrix of cartilage is collagen with an amorphous composite of other proteins and lipids.

16.   Which is a major similarity between bone and cartilage?
   a.   Both have calcium phosphate in their matrices.
   b.   Both have collagen in their matrices.
   c.   Both have osteocytes embedded within a matrix.
   d.   Both have similar compressional strength.
   e.   Both have chitin in their matrices.

17.   Which is NOT part of the axial skeleton?
   a.   sternum
   b.   ribs
   c.   vertebrae
   d.   metatarsals
   e.   cranium

18.   Which is NOT part of either the pelvic or pectoral girdle?
   a.   scapula
   b.   clavicle
   c.   carpal
   d.   ilium
   e.   sacrum

19.   The portion of a myofibril from one Z line to the next is called
   a.   a muscle fiber.
   b.   an H zone.
   c.   a sarcomere.
   d.   a sarcolemma.
   e.   a thin filament.

20.   The characteristic banding pattern of skeletal muscle is the consequence of overlapping
   a.   myosin filaments.
   b.   actin filaments.
   c.   thin filaments.
   d.   actin and myosin filaments.
   e.   pairs of thick filaments.

21.   What reaction would injecting calcium ions into a muscle fiber illicit?
   a.   The fiber would contract.
   b.   The fiber would not contract unless stimulated by a large number of motor neurons.
   c.   The fiber would not contract unless the chemical added was removed.
   d.   Z lines would move further apart.
   e.   The transverse tubules would close, and an action potential would remain at the surface of the fiber.

22. What would happen if a muscle fiber was injected with a protein known to cover and inactivate myosin heads?
   a. The fiber would contract.
   b. The fiber would not contract unless stimulated by a large number of motor neurons.
   c. The fiber would not contract unless the chemical added was removed.
   d. Z lines would move further apart.
   e. The transverse tubules would close, and an action potential would remain at the surface of the fiber.

23. Which accurately compares skeletal and smooth muscle?
   a. Both have obvious striations.
   b. Both have one nucleus per muscle fiber.
   c. Both contract by means of actin and myosin fibers.
   d. Skeletal muscle fibers are seldom anchored to bone, while smooth muscle usually is.
   e. Skeletal muscle has a more disorganized arrangement of actin and myosin than does smooth muscle.

24. In what way are smooth and cardiac muscle similar?
   a. Both have obvious striations.
   b. Both are activated by the visceral sensory nervous system.
   c. Both are activated by the autonomic nervous system.
   d. Both are multinucleate.
   e. Both have intercalated disks.

Concept Map Construction

Construct a concept map for each group of terms. Be sure to include appropriate connector phrases. You may add other terms as necessary and use terms in the plural or singular form.

1. vertebrate skeleton, dermal bone, hammer (middle ear bone), calcitonin, amphibian
2. lung, vertebrate integument, feather, fish scale, diffusion
3. neurotransmitter, sarcomere, ATPase, sarcoplasmic reticulum, mitochondrion

Chapter 27

Processing Food And Providing Nutrition: The Digestive System

Section Concept Map

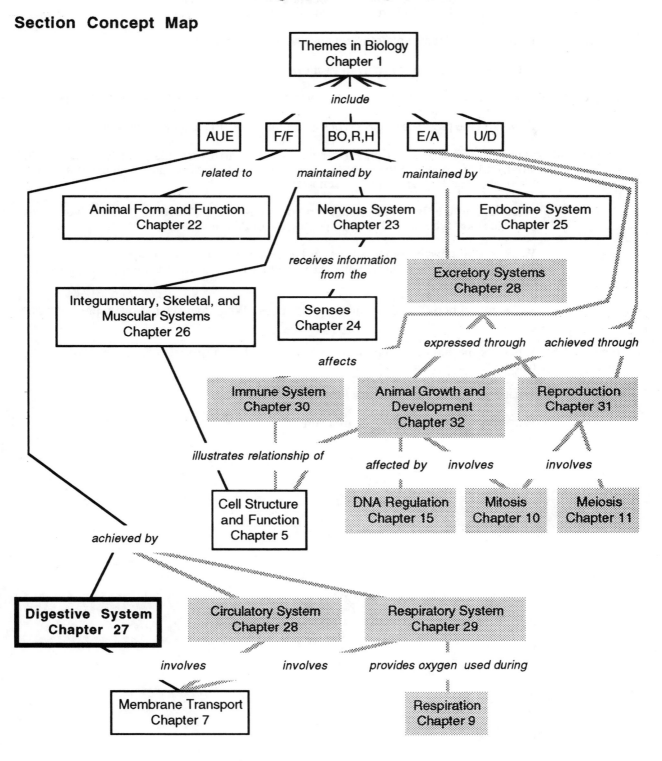

Go Figure!

Figure 27-1:   Which branch of the peripheral nervous system is involved in coordinating contraction of the intestines' muscular wall? What tissue composes the luminal edge of the intestinal mucosa?

Are salivary glands exocrine or endocrine glands? How many salivary glands are there all together? Remember this figure shows only the right side.

Figure 27-2:   Identify the pharynx, epiglottis, tongue and trachea in this figure.

What role does the tongue play in swallowing?

Figure 27-3:   Based on their location, parietal, mucous and chief cells are part of what tissue type?

How does the presence of gastric pits affect the total surface area of the stomach's walls?

Figure 27-4:   Which branch of the peripheral nervous system carries signals from the brain to the stomach in response to the sight or smell of food?...carries signals from the stomach to the brain when food arrives at the stomach?

According to the figure, the stomach produces gastrin. What organ system would you place the stomach in based upon its production of this substance?

Figure 27-5:   What type of muscle comprises the cardiac sphincter?...the pyloric sphincter? To act as an effective one way value, what do you think is the normal state of these muscles: contracted or relaxed?

At the top right of the illustration are two tubes. What does the red one represent?...the gray one?

Figure 27-6:   Where is the lumen of the small intestine in relation to the illustration on the right, towards the top or the bottom?

What type of tissue is immediately below the mucosa?

What tissue type forms the wall of a villus?

Matching Questions

Write the letter of the phrase that best matches the numbered term on the left. Use each only once.

_____ 1.   digestion

_____ 2.   mouth

_____ 3.   mucosa

_____ 4.   serosa

_____ 5.   mucin

_____ 6.   pharynx

_____ 7.   gastric juice

_____ 8.   hydrochloric acid

_____ 9.   parietal cells

_____10.   glucoreceptors

_____11.   pyloric sphincter

_____12.   cardiac sphincter

_____13.   alcohol

_____14.   trypsin

_____15.   secretin

_____16.   microvilli

_____17.   lymphatic vessels

_____18.   lactase

_____19.   incomplete digestive tract

_____20.   filter feeder

a.   a major protein lubricant

b.   a specific nonprotein component of gastric juice

c.   chemical and physical breakdown of food

d.   hormone produced by endocrine cells in the wall of the small intestine

e.   one of few foods absorbed in the stomach

f.   glandular inner lining of the digestive tract

g.   cytoplasmic extensions which act to increase the surface area of the small intestine

h.   muscular valve between the stomach and small intestine

i.   proteolytic enzyme of the pancreas

j.   connective and epithelial tissue covering the digestive tract

k.   does not have a mouth and an anus

l.   collectively, all of the stomach's secretions

m.   found in the hypothalamus

n.   muscular valve between the stomach and esophagus

o.   obtains food suspended in water

p.   back wall of the oral cavity

q.   enzyme found in some epithelial plasma membranes

r.   involved in fat transport from the small intestine

s.   produce acid component of gastric juice

t.   site of the initiation of starch digestion

Multiple Choice Questions

1. Food in the stomach next enters the
    a. pancreas.
    b. liver.
    c. esophagus.
    d. small intestine.
    e. large intestine.

2. Food in the mouth next enters the
    a. pancreas.
    b. liver.
    c. esophagus.
    d. small intestine.
    e. large intestine.

3. Which is the longest segment of the digestive tube?
    a. esophagus
    b. stomach
    c. colon
    d. small intestine
    e. large intestine

4. Of those listed, which is the shortest segment of the digestive tube?
    a. esophagus
    b. stomach
    c. colon
    d. small intestine
    e. large intestine

5. Which best describes filter feeders?
    a. Food particles may be trapped in mucus or strained from water.
    b. Digestion must be intracellular because the food particles are large.
    c. They possess extra stomach chambers.
    d. They are generally discontinuous feeders.
    e. They possess a cecum.

6. Which best describes ruminant animals?
    a. Food particles may be trapped in mucus or strained from water.
    b. Digestion must be intracellular because the food particles are large.
    c. They possess extra stomach chambers.
    d. They are generally continuous feeders.
    e. They possess a cecum.

7. Which IS NOT a function associated with the mouth and oral cavity?
    a. Mechanical digestion of food begins here.
    b. Chemical digestion of food is begun as starch is hydrolyzed.
    c. Food is mixed with a lubricating mucus to form a bolus.
    d. Lysozymes digest protein in the food.
    e. Salivary gland secretions are added here.

8. Which IS NOT a function associated with the stomach?
    a. Mechanical digestion of food continues here.
    b. Chemical digestion continues as bile is added to the food.
    c. Hydrochloric acid is added to food.
    d. The acidic environment kills most bacteria on the food.
    e. Proteins are denatured and digested.

9. Which describes chyme entering the large intestine?
   a. It has a pH of 8.2.
   b. It is very watery.
   c. It contains many nutrients which will be absorbed by the colon.
   d. It contains a great deal of pepsinogen from the stomach.
   e. It contains secretin.

10. Which describes chyme entering the large intestine?
    a. It has an acidic pH.
    b. It is very dehydrated.
    c. Most of the nutrients have already been absorbed.
    d. It contains a great deal of trypsin from the stomach.
    e. It contains cholecystokinin.

11. What do salivary glands and the pancreas have in common?
    a. Both secrete bicarbonate buffer.
    b. Both secrete carbohydrases.
    c. Both empty their secretions into the blood.
    d. Both secrete nucleases.
    e. Both secrete lipases.

12. In what way do salivary glands and the pancreas differ?
    a. Only one of them secretes carbohydrases.
    b. Only one of them is an exocrine gland.
    c. Only one of them has a tube opening directly into the digestive tract.
    d. Only one of them produces mucin.
    e. Only one of them contains epithelial tissue.

13. Some people with a weight problem have the size of their stomach reduced by surgery. How would this help them lose weight?
    a. It reduces the number of mucosal cells producing gastrin, which in turn, reduces the ability of the stomach to absorb nutrients. Thus, a person retains fewer calories for the same amount of food eaten.
    b. It reduces the number of mucosal cells producing chyme, which in turn, reduces the ability of the stomach to absorb nutrients. Thus, a person retains fewer calories for the same amount of food eaten.
    c. It makes the stomach smaller, so a smaller volume of food stimulates the stretch receptors. This produces a feeling of fullness after less food is eaten, and the individual eats less and loses weight.
    d. It makes the stomach smaller, so food enters the duodenum more rapidly. This results in rising blood glucose level, creates a feeling of fullness, and the individual stops eating.
    e. The pyloris of the stomach is removed, so food enters the duodenum more rapidly. This results in earlier high blood glucose levels, and the individual feels full sooner and eats less.

14. Some people with a weight problem have part of their small intestine surgically by-passed. Why would this help them lose weight?
    a. It reduces the number of mucosal cells producing trypsin, which in turn, reduces the ability of the small intestine to absorb nutrients. Thus, a person retains fewer calories for the same amount of food eaten.
    b. It reduces the number of mucosal cells producing chyme, which in turn, reduces the ability of the small intestine to absorb nutrients. Thus, a person retains fewer calories for the same amount of food eaten.
    c. It makes the small intestine smaller, so a smaller volume of food stimulates the stretch receptors. This produces a feeling of fullness after less food is eaten, and the individual eats less and loses weight.
    d. It makes the small intestine smaller, so food comes in contact with a smaller surface area and less of it is absorbed. Thus, a person retains fewer calories for the same amount of food eaten.
    e. The pyloris of the small intestine is removed , so food enters the duodenum more rapidly. This results in earlier high blood glucose levels, and the individual feels full sooner and eats less.

15. Which part of the brain is inhibited when blood glucose levels are high?
    a. the hunger center of the thalamus
    b. the hunger center of the hypothalamus
    c. the hunger center of the pons
    d. the hunger center of the medulla
    e. the hunger center of the cortex

16. Which part of the brain is stimulated by high blood glucose levels?
    a. the hunger center of the somatic sensory cortex
    b. motor centers of the somatic cortex
    c. glucoreceptors
    d. cholecystokinin
    e. somatic sensory receptors

17. Which regulatory signal plays a role in gastric juice secretion once food has entered the stomach?
    a. somatic stimulation from the brain in response to the initial smell of food
    b. somatic stimulation of the stomach from the brain
    c. autonomic stimulation of the stomach from the brain
    d. sympathetic stimulation of the stomach from the brain
    e. the release of secretin from endocrine cells of the stomach

18. Which regulatory signal plays a role in gastric juice secretion once food has entered the stomach?
    a. somatic stimulation from the brain in response to the initial smell of food
    b. somatic sensory stimulation from the brain
    c. sympathetic motor stimulation of the stomach from the brain
    d. sympathetic sensory stimulation of the stomach from the brain
    e. the release of gastrin from endocrine cells of the stomach

19. Which is listed by your text as a product of bacterial metabolism absorbed by the human colon?
    a. vitamin B_1
    b. vitamin K
    c. vitamin C
    d. acetic acid
    e. pyruvate

20. Which is listed by your text as a product of bacterial metabolism absorbed by the human colon?
    a. vitamin B_{12}
    b. vitamin E
    c. vitamin D
    d. folic acid
    e. pyruvate

Concept Map Construction

Construct a concept map for each group of terms. Be sure to include appropriate connector phrases. You may add other terms as necessary and use terms in the plural or singular form.

1.   endocrine gland, saliva, gastrin, glandular epithelium, autonomic nervous system
2.   bolus, esophagus, pharynx, epiglottis, peristalsis
3.   liver, gall bladder, lipase, microvilli, exocrine gland

Chapter 28

Maintaining The Constancy Of The Internal Environment: The Circulatory and Excretory Systems

Section Concept Map

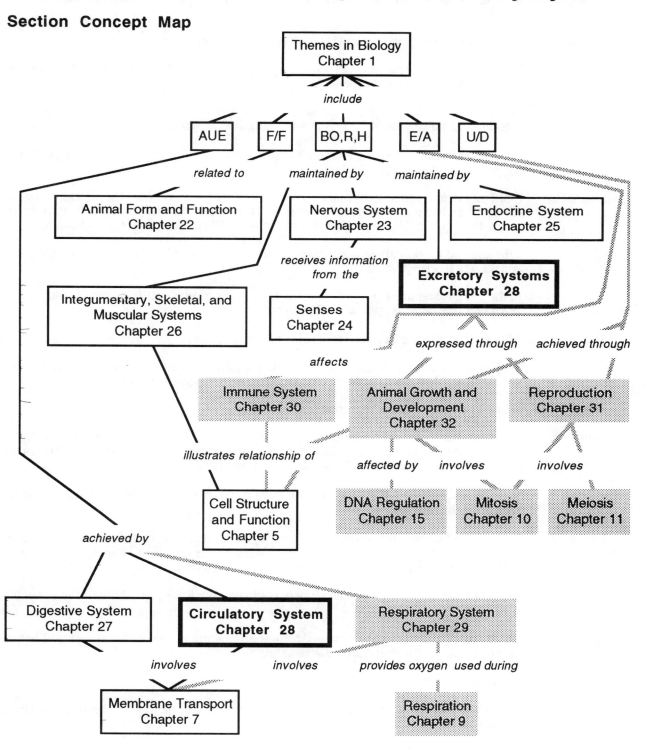

Go Figure!

Figure 28-2:   Name two physical differences between arteries and veins illustrated by this figure.

Which basic tissue type is endothelium?...the elastic layers?

Is there a structural difference between capillaries on the arteriole side of a capillary net and the venule side?

Figure 28-3:   Judging from the drawing, why is blood pressure taken on the upper arm?

Why isn't a sound heard through the stethoscope when the sphygmomanometer cuff has been pumped to 180 mm of mercury?

For a person with a systolic pressure of 120, when the cuff has 120 mm mercury pressure, the pulse sounds are regular but intermittent. Why?

For a person with a diastolic pressure of 80, when the cuff has 80 mm mercury pressure, the pulse sound disappears. Why?

Figure 28-5:   Fluid movement across a capillary is the result of blood pressure on the arteriole side of the capillary bed and osmotic pressure on the venule side. What physical changes occur across the capillary bed to produce this difference? What has caused these changes?

Why are red blood cells entering the arteriole side colored red and those leaving the venule side shaded blue?

In the middle-bottom portion of this drawing, there is a cutaway of the endothelial lining. What does the large, purple object represent?

Why is molecular oxygen (O_2) shown leaving the capillary?

Why is molecular oxygen shown leaving the capillary on the arteriole side, but not the venule side? Why does the reverse relationship hold for carbon dioxide?

Figure 28-6:   What do the gray arrows represent?...the red arrows?

Figure 28-7:   There are valves between ventricles and atria called atrioventricular valves (on the right side the tricuspid valve, and on the left the bicuspid or mitral valve). What would happen to blood flow if these valves did not function properly?

What would happen to the blood flow pattern if blood leaked through the muscle tissue separating the right and left ventricles?

Figure 28-8:   Figure 28-5 indicates that on the arteriole side of a tissue capillary bed, molecular oxygen moves from blood to surrounding tissue. What is the direction of molecular oxygen movement in the capillary beds of the lung?...of carbon dioxide?

How do arteries and veins of the pulmonary circulation differ from those of the systemic circulation?

Figure 28-13: There are three lightning bolt-shaped arrows originating at the sinoatrial node. What is the significance of their lengths relative to one another?

What do the black arrows represent?

Figure 28-17: List several morphological characteristics you may use to distinguish between the five leukocyte types.

Figure 28-18:   Your text describes blood clotting. Which aspects of that description are missing from this diagram? At which points in the diagram do the missing factors impact the cascade of events?

Figure 28-20:   Characterize the "hearts" of a grasshopper or earthworm and compare them to the human heart.

   If these structures are considered functional hearts, what role must they play in circulation?...What tissue must be present?

Figure 28-22:   Lymph is propelled through lymphatic vessels. This requires a force. How is it generated?

   The diagram shows lymph emptying into a vein. What does this suggest about the pressure difference between lymphatic vessels and circulatory system veins?

Figure 28-24:   Trace the path blood travels from renal artery to renal vein. Through how many capillary systems will blood entering the glomerulus pass before reaching the renal vein?

   Fluid passes from glomerulus to Bowman's capsule. What does this suggest about the fluid pressure in each structure?

Figure 28-25:   What do the black arrows illustrate?

   What does the colored arrow next to the word urine have in common with the black arrows?

   How do arrow color differences along tubules relate to functional differences?

   What is the significance of the arrow above the word cortex?

   Along which sections of the tubules are osmotic gradients responsible for the direction of the arrow? Along which sections is pressure responsible for the direction? Along which sections is active transport responsible for the direction?

Matching Questions

Write the letter of the phrase that best matches the numbered term on the left. Use each only once.

_____ 1. interstitial fluid

_____ 2. blood pressure

_____ 3. endothelium

_____ 4. atherosclerosis

_____ 5. vasodilation

_____ 6. vasoconstriction

_____ 7. edema

_____ 8. capillary

_____ 9. varicose vein

_____10. venule

_____11. systemic circuit

_____12. atrioventricular valve

_____13. sinoatrial node

_____14. lymphatic capillary

_____15. osmoregulation

_____16. uric acid

_____17. nephron

_____18. proximal convoluted tubule

_____19. antidiuretic hormone

_____20. ectotherm

a. results from decreased sympathetic nervous system stimulation

b. increased water retention in a tissue

c. specialized cardiac tissue that regulates heart contractions

d. microscopic connecting vessel between artery and vein

e. maintenance of water and salt content

f. blood vessel intermediate in size between capillary and vein

g. the epithelial lining of blood vessels

h. functional kidney unit

i. is found between and surrounding all cells

j. secretion increases with an increase in blood osmotic pressure

k. build up of fatty plaque inside an artery

l. organism in which temperature homeostasis is dependent upon external heat source

m. nontoxic nitrogen compound

n. located between atria and ventricles

o. results from elastic recoil of arteries

p. kidney structure emerging from Bowman's capsule

q. overdistended blood vessel

r. microscopic dead end vessels physically near blood capillaries

s. results from increased sympathetic nervous system stimulation

t. all vessels past the left ventricle and before the right atrium

Multiple Choice Questions

1. Which returns blood to the human heart from the systemic circulation?
    a. arteries
    b. arterioles
    c. capillaries
    d. veins
    e. venules

2. Which returns blood to the human heart from the pulmonary circulation?
    a. arteries
    b. arterioles
    c. capillaries
    d. veins
    e. venules

3. According to your text, most veins in the human systemic circulation are transporting blood
    a. downward, with the force of gravity.
    b. upward, against the force of gravity.
    c. horizontally, neither with nor against the force of gravity.
    d. to the liver.
    e. to the kidneys.

4. Most veins in the human pulmonary circulation are transporting blood
    a. downward, with the force of gravity.
    b. upward, against the force of gravity.
    c. horizontally, neither with nor against the force of gravity.
    d. to the liver.
    e. to the kidneys.

5. Which valve separates the right atrium and right ventricle of the human heart?
    a. bicuspid valve
    b. tricuspid valve
    c. mitral valve
    d. right semilunar valve
    e. left semilunar valve

6. Which valve separates the right ventricle and pulmonary artery of the human heart?
    a. bicuspid valve
    b. tricuspid valve
    c. mitral valve
    d. right semilunar valve
    e. left semilunar valve

7. Which is most likely to occur in intestinal arterioles during times of emotional or physical stress?
    a. vasodilation
    b. vasoconstriction
    c. decreased blood pressure
    d. lower resistance to blood flow
    e. total vascular shut down

8. Which is most likely to occur in skeletal muscle arterioles during times of emotional or physical stress?
    a. vasodilation
    b. vasoconstriction
    c. increased blood pressure
    d. higher resistance to blood flow
    e. total vascular shut down

9. In which sequence does blood pass through the chambers of the heart?
   a. right atrium, left atrium, right ventricle, and left ventricle
   b. right atrium, right ventricle, left atrium, and left ventricle
   c. left atrium, right atrium, left ventricle, and right ventricle
   d. left atrium, left ventricle, right ventricle, and right atrium
   e. right ventricle, right atrium, left ventricle, and left atrium

10. In which sequence does blood pass through the valves of the heart?
    a. tricuspid, right semilunar, left semilunar, and bicuspid
    b. bicuspid, right semilunar, left semilunar, and tricuspid
    c. tricuspid, right semilunar, bicuspid, and left semilunar
    d. right semilunar, left semilunar, tricuspid, and bicuspid
    e. right semilunar, tricuspid, left semilunar, and bicuspid

11. Ventricular stimulation by the atrioventricular (AV) node results in
    a. a wave of muscular contraction beginning at the AV node and proceeding down the ventricles.
    b. a wave of muscular contraction beginning at the end of the ventricle opposite the AV node and proceeding up the ventricles.
    c. a wave of muscular contraction beginning at the AV node and proceeding down the atria.
    d. a wave of muscular contraction beginning at the end of the atrium opposite the AV node and proceeding up the atria.
    e. a wave of muscular contraction forcing blood through the bicuspid valves.

12. Atrial stimulation by the sinoatrial (SA) node results in
    a. a wave of muscular contraction beginning at the SA node and proceeding across the atria.
    b. a wave of muscular contraction beginning at the end of the atria opposite the SA node and proceeding up each atrium.
    c. a wave of muscular contraction beginning at the AV node and proceeding down the atria.
    d. a wave of muscular contraction beginning at the end of the atrium opposite the AV node and proceeding up the atria.
    e. a wave of muscular contraction forcing blood through the semilunar valves.

13. Vasoconstriction occurs when arteriole smooth muscle responds to norepinephrine released by sympathetic neurons. There are no parasympathetic neurons to the arteriole musculature of most body tissues. Which explanation of how <u>vasodilation</u> may be achieved in these tissues is consistent with this and the material presented in your text?
    a. Release of a hormone may trigger vasodilation.
    b. Smooth muscle relaxes in response to increasing blood O_2 concentration.
    c. Smooth muscle relaxes in response to increasing blood glucose concentrations.
    d. Smooth muscle relaxes in response to decreasing blood levels of lactic acid.
    e. Release of mucin may trigger vasodilation

14. Vasoconstriction occurs when arteriole smooth muscle responds to norepinephrine released by sympathetic neurons. There are no parasympathetic neurons to the arteriole musculature of most body tissues. Which explanation of how <u>vasodilation</u> may be achieved in these tissues is consistent with this and the material presented in your text?
    a. Increasing blood levels of hypothalamic releasing factors may stimulate vasodilation.
    b. Smooth muscle tissue relaxes in response to decreasing blood CO_2 concentration.
    c. Smooth muscle relaxes as a result of decreased sympathetic stimulation.
    d. Decreasing blood pH may stimulate vasodilation.
    e. Increasing blood levels of calcium ions may stimulate vasodilation.

15. Which blood component contains antibodies?
    a. albumin
    b. globulins
    c. erythrocytes
    d. platelets
    e. basophils

16. Which blood component releases chemicals that trigger an inflammatory response?
    a. albumin
    b. globulins
    c. erythrocytes
    d. platelets
    e. basophils

17. In which way is thrombin and a pancreatic protease similar?
    a. Both facilitate the enzymatic breakdown of food.
    b. Both are proteolytic enzymes.
    c. Both are capable of converting enzymes from inactive to active forms.
    d. Both activate Factor VIII.
    e. Both produce an inflammatory response at the site of an injury.

18. In which way is Factor X and trypsin similar?
    a. Both facilitate the enzymatic breakdown of food.
    b. Both are proteolytic enzymes.
    c. Both are capable of converting enzymes from inactive to active forms.
    d. Both activate Factor VIII.
    e. Both produce an inflammatory response at the site of an injury.

19. Which is common to both the pulmonary and systemic circulatory circuits in the frog?
    a. the lung capillaries
    b. the right atrium
    c. the left atrium
    d. the ventricle
    e. the pulmonary vein

20. What advantage does an amphibian three-chambered heart have over a fish two-chambered heart?
    a. Three chambers allows atria to contract slightly before the ventricle, increasing heart efficiency.
    b. Three chambers allows maintenance of blood pressure to gas exchange organs and the general body circulation.
    c. Three chambers allows return of oxygenated blood to the heart assuring a constant oxygen supply to cardiac muscle.
    d. Three chambers allows for higher blood pressure and greater flow rate to gas exchange organs.
    e. Three chambers eliminates mixing of oxygenated and unoxygenated blood.

21. How are the ureters and esophagus similar?
    a. Both are passive tubes through which materials move by gravity.
    b. Both are tubes leading to a passive storage organ.
    c. Both are tubes through which materials are moved by peristalsis.
    d. Both absorb water from their contents and return it to the circulatory system.
    e. Both transport metabolic wastes

22. Which is an accurate comparison of urea and uric acid?
    a. Urea is formed from the metabolic breakdown of proteins, and uric acid from catabolism of nucleic acids.
    b. Urea is formed from the metabolic breakdown of nucleic acids, and uric acid from catabolism of protein.
    c. Both are formed by the kidneys.
    d. Both are relatively nontoxic nitrogenous molecules at low concentrations.
    e. Uric acid is excreted in water, and urea as a dry paste.

23. Filtrate in the proximal convoluted tubule next enters
    a. the descending limb of Henle's loop.
    b. the ascending limb of Henle's loop.
    c. the distal convoluted tubule.
    d. the glomerulus.
    e. Bowman's capsule.

24. Filtrate in the ascending limb of Henle's loop next enters
    a. the descending limb of Henle's loop.
    b. the proximal convoluted tubule.
    c. the distal convoluted tubule.
    d. the glomerulus.
    e. Bowman's capsule.

25. Which describes urine formation in a nephron?
    a. Filtration is a selective process because only wastes enter the filtrate.
    b. Some substances are secreted into the urine by the distal tubules.
    c. Water is actively transported from the filtrate back into the blood.
    d. Sodium ions passively diffuse from the distal tubules into surrounding tissue.
    e. Nutrients passively diffuse from the collecting duct into surrounding tissue.

26. Which describes urine formation in a nephron?
    a. Filtration is a selective process because only water enters the filtrate.
    b. Blood sugar is secreted into the urine by the proximal tubules.
    c. Water is passively transported from the filtrate back into the blood.
    d. Sodium ions passively diffuse from the proximal tubules into surrounding tissue.
    e. Nutrients are actively transported from the collecting duct into surrounding tissue.

Concept Map Construction

Construct a concept map for each group of terms. Be sure to include appropriate connector phrases. You may add other terms as necessary and use terms in the plural or singular form.

1. peripheral nervous system, smooth muscle, capillary, lactic acid, molecular oxygen
2. diffusion gradient, pressure gradient, endocytosis, endothelium, edema
3. hypothalamus, ADH, aldosterone, collecting duct, glomerulus

Chapter 29

Gas Exchange: The Respiratory System

Section Concept Map

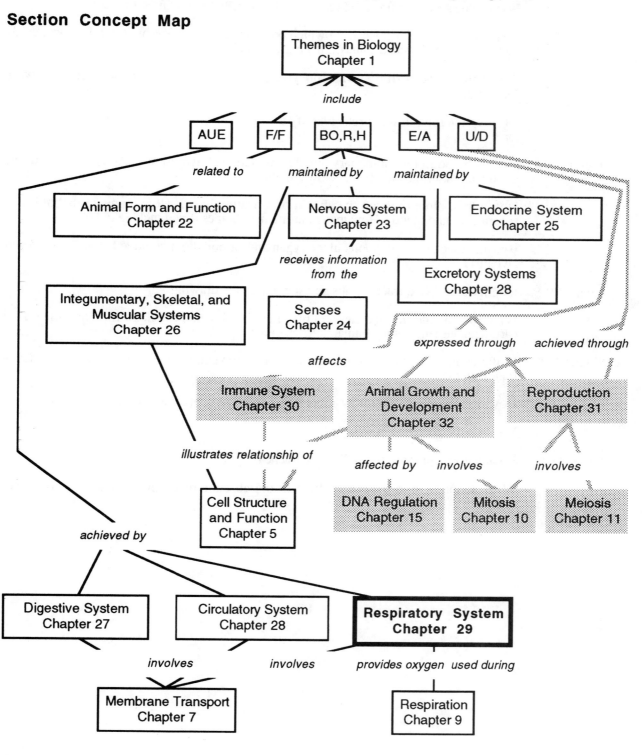

Go Figure!

Figure 29-2:   From highest to lowest, what is the direction of the oxygen concentration gradient in these tissues?...carbon dioxide concentration gradient?

What is the specific name used for the epithelial cells shown lining the capillary?

Figure 29-3:   Observe the vocal cord inset. In which position would air passing over the vocal cords make the lowest pitched sound?...the highest pitched?

Based upon its position, what do you think the literal meaning is for the word "epiglottis?"

Observe the lung tissue inset. Identify a single alveolus. What is the large tube at the top?

Figure 29-4:   Why is blood in the pulmonary venule circulation shown in red? Why is blood in the pulmonary arteriole circulation shown in blue?

Does blood directly touch the alveoli of a healthy lung? How many epithelial layers separate air and blood? How many plasma membranes if you include red blood cells?

By what mechanism do molecular oxygen and carbon dioxide move between blood and alveolus?

Figure 29-5:   Identify the larynx, trachea, ribs and sternum in the drawing.

What rolls do the ribs and sternum play in inhalation and exhalation? What role do the cartilagenous rings of the human trachea play?

Figure 29-9:   In the countercurrent flow arrangement, does oxygen always move down its concentration gradient?

How does countercurrent flow affect the concentration gradient of carbon dioxide?

What do the percentages refer to?

Figure 29-11: What tissues comprise the wall of the trachea?

What adaptive significance is there to the absence of cartilagenous rings in insect tracheae?

Matching Questions

Write the letter of the phrase that best matches the numbered term on the left. Use each only once.

_____ 1. aerobic respiration

_____ 2. respiratory surface

_____ 3. cutaneous

_____ 4. nostril

_____ 5. glottis

_____ 6. bronchus

_____ 7. alveolus

_____ 8. capillary

_____ 9. erythrocyte

_____10. pleurisy

_____11. ventilation

_____12. external intercostals

_____13. oxyhemoglobin

_____14. carbaminohemoglobin

_____15. bicarbonate ion

_____16. carbonic anhydrase

_____17. medulla

_____18. central chemoreceptors

_____19. spiracle

_____20. pharyngeal "basket"

a. air pocket at the end of a bronchiole

b. opening into the larynx

c. breathing

d. microscopic vessel connecting artery and vein

e. catalyzes the formation of carbonic acid

f. any thin, moist epithelium across which gases readily diffuse

g. inflammation of the double-membraned sac surrounding the lungs

h. CNS hydrogen ion receptors

i. $Hb(CO_2)$

j. external opening into the human respiratory tract

k. requires an inorganic final electron acceptor

l. location of the respiratory center

m. a filter feeding organ which participates in gas exchange

n. aid in ventilation

o. opening into an insect trachea

p. pertaining to the skin

q. HCO_3^-

r. $Hb(O_2)_4$

s. subdivision of the human trachea

t. anucleated cellular component of blood

Multiple Choice Questions

1. Which is a property of cellular respiratory membranes?
   a. They must be covered with a layer of moisture so that gases may dissolve in water.
   b. They are covered with keratin.
   c. They must be dry so gases will diffuse across them.
   d. They are actively endocytotic.
   e. They transport gases by facilitated diffusion.

2. Which is a property of cellular respiratory membranes?
   a. They must be covered with a layer of collagen so that gases may dissolve before diffusing across.
   b. They are covered with chitin.
   c. They must be covered with mucus so gases will diffuse across them.
   d. They are composed of thin, gas permeable epithelium.
   e. They exchange gases by active transport.

3. Which organism was listed by your text as being a cutaneous respirer?
   a. jellyfish
   b. shark
   c. elephant
   d. dog
   e. lizard

4. Which organism was listed by your text as being a cutaneous respirer?
   a. mosquito
   b. ant
   c. frog
   d. snake
   e. bird

5. What is the direction of the O_2 concentration gradient (from highest to lowest) in an alveolus and surrounding tissues?
   a. from alveolus to bronchiole
   b. from red blood cell to alveolus
   c. from alveolus to blood plasma
   d. from blood plasma to red blood cell
   e. from erythrocyte to alveolus

6. What is the direction of the O_2 concentration gradient (from highest to lowest) in a systemic capillary and surrounding tissues?
   a. from tissue to blood plasma
   b. from blood plasma to red blood cell
   c. from red blood cell to blood plasma
   d. from blood plasma to erythrocyte
   e. from tissue to erythrocyte

7. Which is the correct sequence of human respiratory structures through which <u>inhaled</u> air passes?
   a. nasal cavity, trachea, glottis, alveoli
   b. nasal cavity, glottis, pharynx, alveoli
   c. alveoli, bronchioles, trachea, glottis, nasal cavity
   d. nasal cavity, glottis, larynx, bronchi
   e. nostrils, bronchi, larynx, nasal cavity

8.   Which is the correct sequence of human respiratory structures through which <u>exhaled</u> air passes?
     a.   nasal cavity, trachea, glottis, alveoli
     b.   nasal cavity, glottis, pharynx, alveoli
     c.   alveoli, bronchioles, trachea, glottis, nasal cavity
     d.   nasal cavity, glottis, larynx, bronchi
     e.   nostrils, bronchi, larynx, nasal cavity

Use the following information for questions 9 and 10.

Two major requirements for gas exchange include sufficient surface area across which gases move and a thin respiratory membrane. This structural arrangement allows for the homeostatic maintenance of internal CO_2 and O_2 concentrations. This large, moist surface area also presents a threat to homeostasis in terrestrial animals and aquatic animals.

9.   What homeostatically maintained condition is threatened directly by the respiratory membranes of a terrestrial animal?
     a.   water content of the body
     b.   blood glucose level
     c.   vasodilation
     d.   blood calcium ion level
     e.   neuron resting potential

10.   What homeostatically maintained condition is threatened directly by the respiratory membranes of an aquatic animal living in a hypotonic environment?
     a.   trypsin level in the stomach
     b.   liver glycogen content
     c.   sodium content of tissues
     d.   carbonic anhydrase concentration
     e.   oxygen concentration in the alveoli

11.   Which is a function of the nasal passages?
     a.   They filter air before it enters the bronchi.
     b.   The regulate the volume of air reaching the lungs.
     c.   They contract to force air into the lungs.
     d.   They expand to draw air out of the lungs.
     e.   They provide cartilaginous support for the bronchioles.

12.   Which is a function of the nasal passages?
     a.   They vibrate when air is forced over them to amplify sounds produced by the vocal cords.
     b.   They moisten the air before it enters the bronchi.
     c.   They add mucin to dry the air before it enters bronchioles.
     d.   They prevent food or drink from passing into the glottis.
     e.   The provide cartilagenous support for the trachea.

13.   According to your text, how do bronchi and bronchioles compare?
     a.   Bronchi contain muscles to regulate their size. Bronchioles do not.
     b.   Bronchioles contain rings of cartilage. Bronchi do not.
     c.   Bronchioles contain muscles to regulate their size. Bronchi do not.
     d.   Bronchi passage size is regulated by the sympathetic nervous system. Bronchiole passage size is regulated by the parasympathetic system.
     e.   Bronchioles immediately branch from the trachea. Bronchi branch from bronchioles.

14. According to your text, how do bronchi and bronchioles compare?
    a. Bronchi contain muscles to regulate their size and so do bronchioles.
    b. Bronchi contain rings of cartilage. Bronchioles do not.
    c. Neither bronchi nor bronchioles contain muscles to regulate their size.
    d. Bronchi passage size is regulated by the parasympathetic nervous system. Bronchiole passage size is regulated by the sympathetic system.
    e. Bronchioles immediately branch from the trachea. Bronchi branch from alveoli.

15. Immediately after the diaphragm and external intercostals contract, where is air pressure the lowest?
    a. external air
    b. nasal passages
    c. pharynx
    d. trachea
    e. bronchi

16. Immediately after the diaphragm and external intercostals contract, where is air pressure the highest?
    a. external air
    b. nasal passages
    c. pharynx
    d. trachea
    e. bronchi

17. Which blood component transports the largest quantity of oxygen?
    a. plasma
    b. red blood cell
    c. platelet
    d. carbonic anhydrase
    e. carbaminohemoglobin

18. Which blood component transports the largest quantity of carbon dioxide?
    a. plasma
    b. red blood cell
    c. platelet
    d. carbonic anhydrase
    e. carbaminohemoglobin

19. What comprises oxyhemoglobin?
    a. plasma and oxygen
    b. oxygen and carbonic anhydrase
    c. carbon dioxide and hemoglobin
    d. oxygen and hemoglobin
    e. hemoglobin and carbonic anhydrase

20. What comprises carbaminohemoglobin?
    a. plasma and oxygen
    b. oxygen and carbonic anhydrase
    c. carbon dioxide and hemoglobin
    d. oxygen and hemoglobin
    e. hemoglobin and carbonic anhydrase

21. Where is the human respiratory center located?
    a. spinal cord
    b. pons
    c. medulla
    d. hypothalamus
    e. cerebral cortex

22. Where are some of the peripheral chemoreceptors located which send sensory information to the human respiratory center?
    a. spinal ganglia
    b. tricuspid valve
    c. atrioventricular node
    d. walls of the aorta
    e. alveoli of the lungs

23. The lamellae of gill filaments serve to
    a. increase surface area for gas exchange.
    b. direct blood flow perpendicular to the flow of water over the gills.
    c. cover and protect delicate gill membranes.
    d. channel blood flow in the same direction as water flow over the gills.
    e. establish a large chamber where oxygen-poor blood can pool.

24. Each insect spiracle leads directly into
    a. a trachea.
    b. a tracheole.
    c. a muscular diaphragm.
    d. a bronchiole.
    e. a lamella.

Concept Map Construction

Construct a concept map for each group of terms. Be sure to include appropriate connector phrases. You may add other terms as necessary and use terms in the plural or singular form.

1. respiratory surface, diffusion, concentration gradient, alveolus, circulatory system
2. red blood cell, hemoglobin, lactic acid, electron transport chain, endothelium
3. air pressure, diaphragm, rib, trachea, epiglottis

Chapter 30

Internal Defense: The Immune System

Section Concept Map

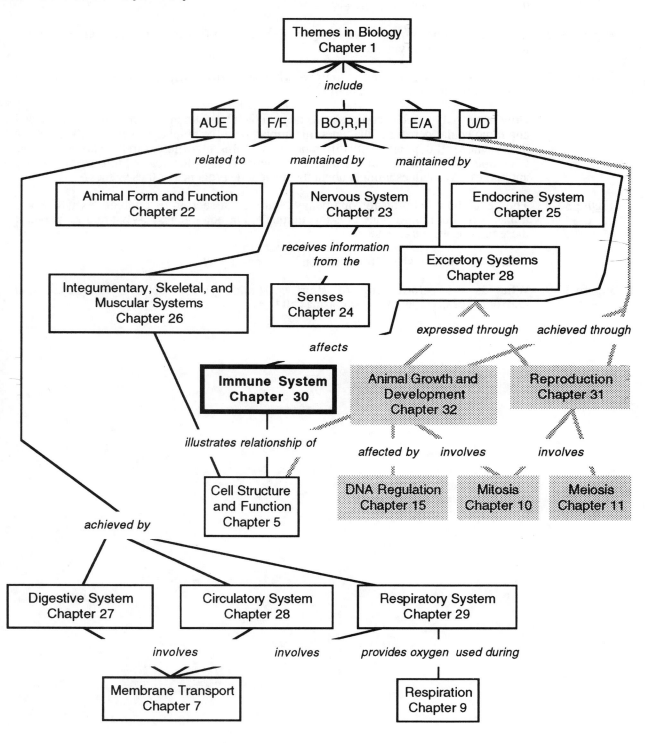

Go Figure

Figure 30-1:   What general type of cellular transport are these cells involved in?

Once inside, what organelle provides enzymes to destroy the bacteria?

Figure 30-2:   Leukocytes are able to squeeze between endothelial cells. What else will be able to leave the capillary through these spaces?

Figure 30-3:   In addition to an immune system role, what other functions are performed by bone marrow?...spleen?...lymphatic tissue?

Figure 30-7:   If an antibody is able to bind an antigen, what, aside from shape, must be complimentary between immunoglobulin and antigen?

How many polypeptides compose one heavy chain? Based upon what you have learned about molecular genetics, what does this imply about the number of genes coding for one heavy chain? If their are many different heavy chains, all different in the variable region, what does this imply about the number of genes coding for heavy proteins of class G antibodies? What does it imply about the nucleotide sequence of these genes?

Figure 30-8:   The gene shown coding for one heavy chain has a section of nucleotides which are transcribed, but not translated. Is this unique to antibody genes? What might happen if these regions were translated?

What is the cellular site of transcription?...of translation?

Figure 30-9:   What is the relationship of stem cell and lymphocyte?....lymphocyte and macrophage?...lymphocyte and plasma cell?

Matching Questions

Write the letter of the phrase that best matches the numbered term on the left. Use each only once.

_____ 1. natural killer cell

_____ 2. pus

_____ 3. interferon

_____ 4. Edward Jenner

_____ 5. smallpox

_____ 6. macrophage

_____ 7. B lymphocyte

_____ 8. T lymphocyte

_____ 9. tears

_____ 10. perforins

_____ 11. interleukin II

_____ 12. histocompatibility antigen

_____ 13. tissue graft

_____ 14. cyclosporine

_____ 15. immunoglobulin

_____ 16. antigen binding site

_____ 17. Susumu Tonegawa

_____ 18. attentuated virus

_____ 19. antigen presentation

_____ 20. immunological recall

a. formed from a large number of leukocytes

b. characterized by the development of elevated blisters on the skin

c. exocrine gland product containing antibodies

d. specific cell-surface antigens

e. first to collect evidence of the two gene-one polypeptide hypothesis

f. may produce plasma cells

g. an antibody

h. a macrophage, B cell interaction

i. nonspecific, lymphocyte-like cell

j. made nonvirulent

k. protein produced by host cells in response to a viral infection

l. released by Helper T lymphocytes

m. fungal compound that suppresses cell-mediated immunity

n. first to use immunization

o. immune system capability made possible by B cells

p. involved in nonphagocytic cell-mediated immunity

q. portion of an antibody complimentary to a foreign protein or polysaccharide

r. proteins released by Cytotoxic T lymphocytes

s. the surgical movement of tissue from one area or one individual to another

t. derived from monocytes

Multiple Choice Questions

1. Which is a component of the human body's nonspecific immune response?
    a. natural killer cells
    b. macrophages
    c. B cells
    d. immunoglobulins
    e. plasma cells

2. Which is a component of the human body's specific immune response?
    a. fever
    b. neutrophils
    c. T cells
    d. complement
    e. interferon

3. Which is the source of antibodies in body fluids?
    a. natural killer cells
    b. macrophages
    c. B cells
    d. immunoglobulins
    e. plasma cells

4. Which group of molecules contains a protein capable of binding to a bacterium and putting holes in its plasma membrane?
    a. immunoglobulins
    b. complement
    c. interferon
    d. antibodies
    e. interleukin II

5. Inflammation is initiated
    a. when leukocytes release complement.
    b. when neutrophils release interferon.
    c. when lymphocytes release chemicals into interstitial fluid.
    d. when damaged tissue cells release chemicals into interstitial fluid.
    e. when killer T cells are damaged.

6. Inflammation includes
    a. protein G binding to a bacterium and putting holes in its plasma membrane.
    b. dilation of local blood vessels.
    c. activation of local tissues cells by interleukin II.
    d. inhibition of vasoconstriction by suppressor T cells.
    e. production of antibodies.

7. Which produces interferon?
    a. natural killer cells
    b. macrophages
    c. B cells
    d. tissue cells
    e. plasma cells

8. Interferon
    a. binds to a bacterium and put holes in its plasma membrane.
    b. attracts phagocytic leukocytes including neutrophils.
    c. blocks a cell's ability to produce viral proteins.
    d. initiates an immune response.
    e. interferes with suppressor T cell functions.

9. Plasma cell antibody formation requires the presence of
   a. complement.
   b. antigens.
   c. basophils.
   d. neutrophils.
   e. eosinophils.

10. Macrophages are produced from
    a. B cells.
    b. monocytes.
    c. Killer T cells.
    d. neutrophils.
    e. basophils.

11. How do B cells and plasma cells compare?
    a. B cells are derived from eosinophils, while plasma cells develop from macrophages.
    b. B cells have a smaller nucleus in proportion to cell size than do plasma cells.
    c. Both B cells and plasma cells produce perforins.
    d. Both are formed within the spleen.
    e. Both produce antibodies.

12. How do B cells and plasma cells compare?
    a. B cells are derived from macrophages, while plasma cells develop from eosinophils.
    b. B cells have a larger nucleus in proportion to cell size than do plasma cells.
    c. B cells produce perforins, while plasma cells do not.
    d. Both are formed within the thymus.
    e. B cells produce antibodies, while plasma cells do not.

13. How do B cells and T cells compare?
    a. B cells are leukocytes, while T cells are lymphocytes.
    b. B cells mature into T cells.
    c. Both are lymphocytes.
    d. Both produce perforins.
    e. B cells are lymphocytes, while T cells are leukocytes.

14. How do B cells and T cells compare?
    a. B cells are lymphocytes, while T cells are leukocytes.
    b. T cells mature into B cells.
    c. Both are eosinophils.
    d. Both produce perforins.
    e. B cells are involved in soluble immune responses, while T cells participate in cell-mediated immunity.

15. Which is a function of T cells?
    a. may participate in antigen production
    b. may participate in perforin production
    c. may participate in interferon production
    d. may participate in complement production
    e. may participate directly in immunoglobulin production

16. Which is a function of T cells?
    a. may participate in antibody production
    b. undergo mitosis to produce plasma cell clones
    c. may participate in histocompatibility antigen production
    d. may participate in heavy chain Ig production
    e. may participate in interleukin II production

17. Commonly, human antibodies are composed of _____ protein molecules?
    a. 1
    b. 2
    c. 3
    d. 4
    e. 5

18. The Ig class of human antibodies typically has _____ heavy chains and _____ light chains.
    a. 1, 1
    b. 2, 2
    c. 1, 2
    d. 2, 1
    e. 3, 3

19. What is unusual about the "gene" coding for the heavy chains of a human Ig antibody?
    a. It actually contains the nucleotide sequence for both heavy and light chains.
    b. It contains the nucleotide sequence to make several different light chains.
    c. It is actually two separate genes.
    d. It runs the length of one whole chromosome.
    e. It is always inherited from the mother.

20. What is unusual about the "gene" coding for the light chains of a human Ig antibody?
    a. It actually contains the nucleotide sequence for both heavy and light chains.
    b. It contains the nucleotide sequence to make several different light chains.
    c. It is actually two separate genes.
    d. It runs the length of one whole chromosome.
    e. It is always inherited from the mother.

Concept Map Construction

Construct a concept map for each group of terms. Be sure to include appropriate connector phrases. You may add other terms as necessary and use terms in the plural or singular form.

1. immune system, circulatory system, digestive system, lymphatic system, skeletal system
2. leukocyte, lymphocyte, phagocytic cell, thymus, platelet
3. transcription, intron, variable region, constant region, protein

Chapter 31

Generating Offspring: The Reproductive System

Section Concept Map

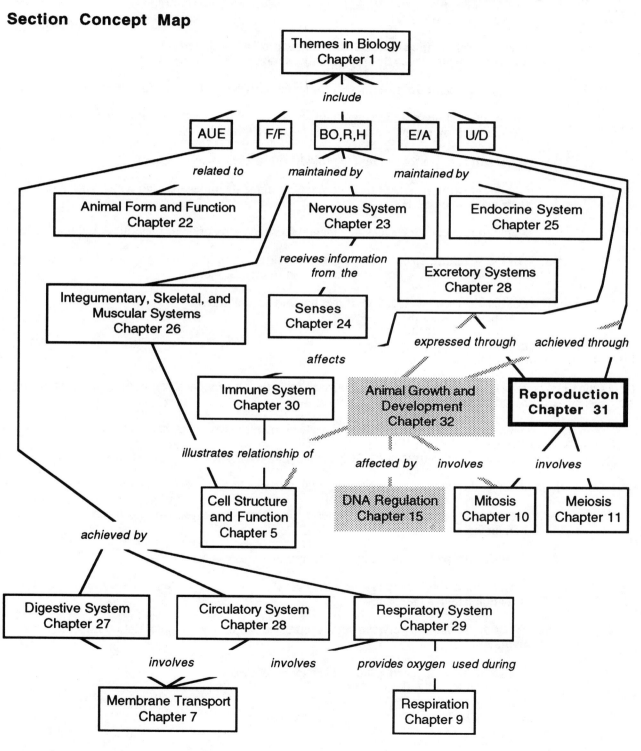

Go Figure!

Figure 31-1:   What kind of cell division is involved in asexual reproduction in multicellular animals?

Some types of parthenogenesis involve only mitosis and some involve mitosis, meiosis and fusion of two cells. Which type most resembles asexual reproduction?...resembles sexual reproduction?

Figure 31-2:   Does external fertilization seem like an appropriate strategy for terrestrial animals? What are some advantages to this reproductive strategy?...some disadvantages?

What are some of the advantages of internal fertilization?...some disadvantages?

Since these slugs produce eggs and sperm, the easiest way to reproduce would be to fertilize their own eggs. Is there a disadvantage to this?

Figure 31-4:   What structure passes through the middle of the corpus spongiosum?

Beginning with the site of their production, name the tubes through which sperm pass on their way out of the body.

Name in sequence the accessory glands sperm pass by on their way out of the body.

Figure 31-5:   On the scanning electron micrograph, identify spermatogonia, primary spermatocytes, secondary spermatocytes, spermatids, and immature sperm.

For each cell type listed, state if it is haploid or diploid.
a.   spermatogonium
b.   primary spermatocyte
c.   secondary spermatocyte
d.   spermatid

Figure 31-7:   Where are sensors located important in negative feedback regulation of human male blood testosterone level?

What is the control unit for human male blood testosterone level?

What acts as the effector for human male blood testosterone level? (See Chapter 22)

Answer the same questions for the proposed mechanism of FSH regulation.

Figure 31-8:   Describe the path followed by sperm in reaching the typical point of fertilization in a human female?

Describe the path followed by a baby during a noncesarean birth.

Based upon this drawing, why do you think so many women experience frequent urination late in a pregnancy?

Figure 31-9:   What do the arrows between stages represent? Identify the potential ovum in each pre-ovulation stage.

Figure 31-10:   For each cell type listed, state if it is haploid or diploid.
a.   oogonium (plural = oogonia)
b.   oocyte
c.   polar body

Figure 31-11:   What do the relative lengths of the red and blue arrows indicate?...relative widths?

What are the relative levels of FSH, LH, estrogen, and progesterone on Day 1 of the menstrual cycle?...Day 7?...Day 14?...Day 21?...Day 28?

What do the blue dashed lines indicate?...the red dashed lines?

Matching Questions

Write the letter of the phrase that best matches the numbered term on the left. Use each only once.

_____ 1.   asexual reproduction

_____ 2.   gonads

_____ 3.   parthenogenesis

_____ 4.   fertilization

_____ 5.   hermaphrodite

_____ 6.   ovum

_____ 7.   ova

_____ 8.   male urethra

_____ 9.   glans

_____ 10.   interstitial cells

_____ 11.   spermatids

_____ 12.   epididymis

_____ 13.   acrosome

_____ 14.   gonadotropins

_____ 15.   secondary sex characteristics

_____ 16.   vulva

_____ 17.   womb

_____ 18.   oogonium

_____ 19.   endometrium

_____ 20.   menopause

a.   an individual that produces both sperm and egg

b.   found within the corpus spongiosum

c.   male haploid cells

d.   general term referring to reproduction in the absence of fertilization

e.   female external genitals

f.   more than one egg

g.   cessation of a woman's reproductive cycle

h.   inner lining of the uterus

i.   the development of a haploid egg into an adult without fertilization

j.   found between the seminiferous tubules

k.   sperm component containing digestive enzymes

l.   premeiotic germ cell in a female

m.   the primary reproductive organs

n.   uterus

o.   one egg

p.   include enlargement of the larynx in a male

q.   fusion of two different cell types during reproduction

r.   hormones with testicular interstitial cells as targets

s.   sits atop a testis

t.   the tip of the penis

Multiple Choice Questions

1. Which describes asexual reproduction?
    a. It introduces great variability to a population by mutation.
    b. It is the parthenogenetic development of haploid cells.
    c. It does not involve fertilization.
    d. It involves fertilization
    e. It involves the fusion of two diploid cells.

2. Which describes sexual reproduction?
    a. It introduces great variability to a population by mutation.
    b. It involves parthenogenetic development of haploid cells.
    c. It does not involve fertilization.
    d. It involves fertilization
    e. It involves the fusion of two diploid cells.

3. Asexual reproduction differs from sexual reproduction in that
    a. asexual reproduction involves the fusion of two cell types, sexual reproduction does not.
    b. asexual reproduction does not require that a diploid organism undergo meiosis to maintain the diploid number of chromosomes in the next generation, sexual reproduction does.
    c. sexual reproduction produces clones of the parent, asexual reproduction does not.
    d. asexual reproduction requires the production of gametes, sexual reproduction does not.
    e. fission is a form of asexual reproduction, while budding is a form of sexual reproduction.

4. Asexual and sexual reproduction are similar in that
    a. both involve gametes.
    b. both produce variability in the offspring by genetic recombination.
    c. both result in the passing of chromosomes from one generation to the next.
    d. both require the production of haploid cells.
    e. both require the production of diploid cells.

5. Some hermaphroditic animals are capable of self-fertilization. The offspring of such an animal would
    a. be clones of the original parent.
    b. have been produced by parthenogenesis.
    c. be different from one another, but probably not as different as offspring produced by cross-fertilization.
    d. probably be more variable in their traits than if they had been produced by cross-fertilization.
    e. have a greater number of cross overs between homologous chromosomes.

6. If parthenogenesis occurs by fusion of a polar body with an ovum, the offspring of such an animal would
    a. be clones of the original parent.
    b. have been produced by hermaphroditic self-fertilization.
    c. be different from one another, but probably not as different as offspring produced by cross-fertilization.
    d. probably be more variable in their traits than if they had been produced by cross-fertilization.
    e. have a greater number of cross overs occur between homologous chromosomes.

7. Why is it important that only a single sperm enter an animal's egg?
    a. If more than one sperm entered the egg, the additional cytoplasm would decrease the surface to volume ratio and interfere with the zygote's ability to exchange materials with its environment.
    b. If two sperm entered the egg, there would be three times the diploid number of chromosomes rather than two times the haploid number. Animals generally can not tolerate such changes in chromosome number.
    c. If two sperm entered the egg, there would be three times the haploid number of chromosomes rather than two times the haploid number. Animals generally can not tolerate such changes in chromosome number.
    d. In order to enter the egg, sperm must destroy part of the egg's plasma membrane. If more than a single sperm entered, it would destroy so much of the membrane that the zygote would not survive.
    e. The question makes a false assumption. If more than one sperm entered the egg there would be no harmful effect.

8. Why is it important that polar body nuclei separate from the oocyte nucleus?
    a. If one remained within the oocyte, the additional nucleoplasm would decrease the surface to volume ratio and interfere with the zygote's ability to exchange materials with its environment.
    b. If one remained within the oocyte, and the ovum which developed from the oocyte was fertilized by a single sperm, the zygote would have three times the haploid number of chromosomes. Animals generally can not tolerate such changes in chromosome number.
    c. If one remained within the oocyte, the ovum which developed from the oocyte could not be fertilized by a single sperm. The presence of two sets of maternal chromosomes would result in the entrance of two sperm so that two sets of paternal chromosomes would pair with two sets of maternal chromosomes.
    d. In order to enter the egg, sperm must destroy part of the egg's plasma membrane. If a polar body remained within the oocyte, there would be a double membrane and sperm could not penetrate.
    e. The question makes a false assumption. It is not important that polar bodies separate from the oocyte.

9. The human penis is an organ which
    a. provides a means of achieving fertilization in a terrestrial environment.
    b. provides a storage site for the maturation of sperm.
    c. produces sperm.
    d. contains four cylinders of erectile tissue.
    e. becomes erect when stimulated by sympathetic motor neurons.

10. The human penis is an organ which
    a. becomes erect when parasympathetic motor neurons stimulate the striated muscle lining its arterioles.
    b. provides an avenue of urine excretion from the body.
    c. produces the sex hormone testosterone.
    d. is the only structure of the body without parasympathetic nerve innervation.
    e. becomes erect in response to release of FSH from the anterior pituitary.

11. What would be the consequence of the corpus spongiosum becoming rigid?
    a. The penis would become rigid.
    b. The penis would become flaccid.
    c. Semen would be ejaculated.
    d. Semen would be unable to pass through the urethra.
    e. Parasympathetic motor neurons would cause vasoconstriction of arterioles leading to the corpus spongiosum.

12. What would be the consequence of the corpus spongiosum becoming limp?
   a. The penis would become rigid.
   b. The penis would become flaccid.
   c. Semen would be ejaculated.
   d. Semen would be unable to pass through the urethra.
   e. Parasympathetic motor neurons would cause vasoconstriction of arterioles leading to the corpus spongiosum.

13. What is produced by seminiferous tubules?
   a. sperm
   b. testosterone
   c. FSH
   d. digestive enzymes
   e. ejaculatory fluid

14. What is produced by interstitial cells of the testes?
   a. sperm
   b. testosterone
   c. FSH
   d. digestive enzymes
   e. ejaculatory fluid

15. Secondary spermatocytes divide to produce
   a. spermatogonia.
   b. primary spermatocytes.
   c. spermatids.
   d. sperm.
   e. interstitial cells.

16. Oogonia divide to produce
   a. oocytes.
   b. the first polar body.
   c. the second polar body.
   d. cells of the corpus luteum.
   e. cells of the follicle.

17. Which is a function of ejaculatory fluid?
   a. It inhibits the hypothalamus from producing and releasing GnRH.
   b. It inhibits the hypothalamus from producing and releasing FSH.
   c. It inhibits the anterior pituitary from releasing LH.
   d. It temporarily neutralizes vaginal acidity.
   e. It temporarily neutralizes the effects of LH.

18. Which is true of ejaculatory fluid?
   a. It stimulates the hypothalamus to produce and release GnRH.
   b. It stimulates the hypothalamus to produce and release FSH.
   c. It provides fructose, an energy source for sperm motility.
   d. It temporarily neutralizes vaginal alkalinity.
   e. It comprises about ten percent of the material released during ejaculation.

19. Which is a function of oviducts?
   a. They are the site of zygote implantation.
   b. They are the site of oogonia development and maturation.
   c. They are they normal site of fertilization.
   d. They are the birth canal.
   e. They are the site of follicle development.

20. Which is a function of the follicle?
    a. It produces testosterone.
    b. It produces estrogen.
    c. It produces FSH.
    d. It is the normal site of fertilization.
    e. It is the location of the endometrium.

21. What is the condition of the endometrium on Day 1 of the menstrual cycle?
    a. It is growing thicker.
    b. It is breaking down.
    c. It is beginning to secrete estrogen.
    d. It is beginning to secrete progesterone.
    e. It is inhibiting production and release of FSH.

22. What is the condition of the follicle on day 14 of the menstrual cycle?
    a. It is beginning to increase in size.
    b. It is beginning to decrease in size.
    c. It has, or is about to rupture and release a maturing oocyte.
    d. It is beginning to secrete estrogen.
    e. It is stimulating the production and release of FSH.

23. Which is the most effective contraceptive?
    a. tubal ligation
    b. female birth control pills
    c. IUD
    d. diaphragm used with a spermicide
    e. sponge

24. Which is the most effective contraceptive method?
    a. male condom
    b. condom used with a spermicide
    c. rhythm method
    d. vasectomy
    e. rhythm method

Concept Map Construction

Construct a concept map for each group of terms. Be sure to include appropriate connector phrases. You may add other terms as necessary and use terms in the plural or singular form.

1. chromosome, meiosis, external fertilization, hermaphrodite, penis
2. mitosis, spermatogenesis, FSH, hypothalamus, crossing over
3. birth control pill, FSH, follicle, oocyte, mitosis

Chapter 32

Animal Growth And Development: Acquiring Form and Function

Section Concept Map

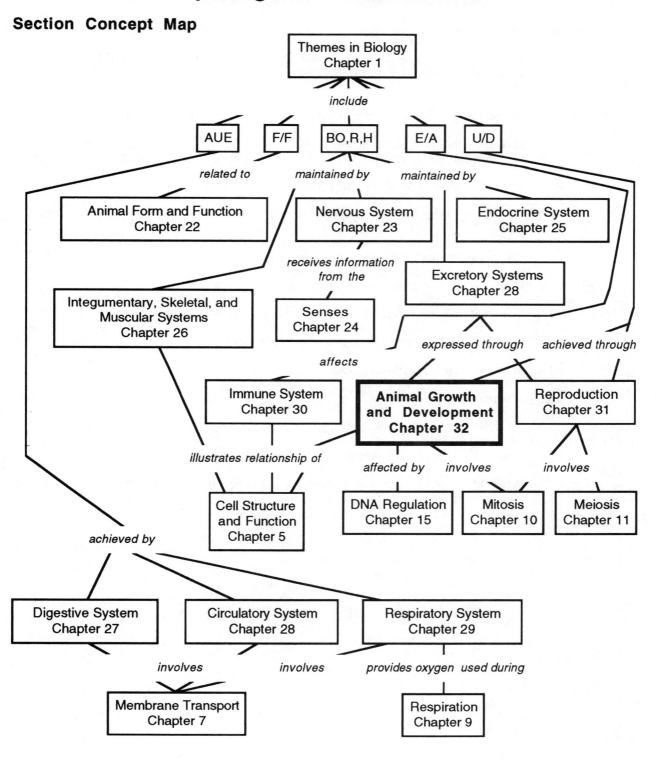

Go Figure!

Figure 32-2:   On the micrograph, identify the acrosome (See Chapter 31) and flagellum.

What role does the acrosome play in fertilization?

How many chromosomes are in a normal human sperm?

Figure 32-4:   Are the cells haploid or diploid?

Is the furrow forming in the photomicrograph (top row, middle column) a mitotic event? What specific cellular component causes this furrow? What are the normal consequences of its formation?

Figure 32-5:   Are different size blastomeres distributed symmetrically in an embryo?

How does the position of the largest sea urchin blastula cells compare to the position of the largest frog blastula cells? How does the position of yolk in each blastula compare?

Figure 32-7:   What does the orange-shaded area represent?

How many germ layers are present in the blastula?...in the early gastrula?...in the late gastrula?

What does "meso" in the term mesoderm mean? How is this an appropriate name?

Figure 32-8:   If this drawing is of frog development, what stage is it in?

Figure 32-13:   Which germ layer is represented by green?...by blue?

How many germ layers comprise the embryonic eye?

Figure 32-16:   Where does fertilization take place?

Where does cleavage begin?

To which germ layer does the trophoblast correspond?

From which group of cells will all three germ layers arise?

What do the green arrows indicate in the upper part of this drawing? What purpose do they serve in the bottom four drawings?

Matching Questions

Write the letter of the phrase that best matches the numbered term on the left. Use each only once.

_____ 1.   larva

_____ 2.   blastomere

_____ 3.   blastodisk

_____ 4.   bithorax

_____ 5.   homeotic gene

_____ 6.   egg activation

_____ 7.   extracellular membrane

_____ 8.   gray crescent

_____ 9.   germ layer

_____10.   ectoderm

_____11.   chordamesoderm

_____12.   primary organizer

_____13.   organogenesis

_____14.   archenteron

_____15.   dorsal lip

_____16.   optic vesicle

_____17.   cornea

_____18.   villi

_____19.   extraembryonic membrane

_____20.   amniotic fluid

a.   a fruit fly mutant with two wings on the last thorax segment

b.   cytoplasmic landmark on the dorsal surface of a frog zygote

c.   cell mitotically derived from the zygote

d.   germ layer which forms the nervous system

e.   space which will give rise to the digestive system lumen

f.   tissue which forms the notochord

g.   marked by an increase in oxygen consumption

h.   branching projections from the chorion

i.   self-feeding immature form

j.   transparent covering of the eyeball

k.   includes the chorion and allantois

l.   part of morphogenesis

m.   formed by materials expelled from the egg cytoplasm

n.   suspends and cushions the developing mammalian embryo

o.   outpocketing of the embryonic brain

p.   area containing early embryonic cells in a bird's egg

q.   a fold of tissue above the blastopore

r.   chordamesoderm

s.   includes the mesoderm

t.   affects the spatial arrangement of the body parts

Multiple Choice Questions

1. All cells of a developing embryo contain the same genetic information because each is produced by
   a. metamorphosis.
   b. organogenesis.
   c. mitosis.
   d. morphogenesis.
   e. differentiation.

2. In addition to selective expression of embryonic genes, maternally derived _____ regulate the expression of development.
   a. tRNAs
   b. rRNAs
   c. mRNAs
   d. dsDNAs
   e. ssDNAs

3. Most animal embryos contain a store of food called yolk. It includes
   a. monosaccharides.
   b. disaccharides.
   c. polysaccharides.
   d. dipeptides.
   e. glycerol.

4. Most animal embryos contain a store of food called yolk. It includes
   a. glycerol.
   b. fatty acids.
   c. tripeptides.
   d. protein.
   e. nucleic acids.

5. Which is a primary sperm function?
   a. form an extracellular membrane around the egg
   b. reduce the cytoplasmic content of the egg
   c. introduce Ca^{++} ions into the egg
   d. contribute a homologous chromosome set
   e. initiate meiosis in the egg

6. Which is a primary sperm function?
   a. form an intracellular membrane in the egg
   b. increase egg's cytoplasmic content
   c. reduce Ca^{++} ion concentration in the egg
   d. reduce the egg's mRNA content
   e. egg activation

7. Which is a consequence of egg activation?
   a. contribution of a homologous chromosome set
   b. formation of an extracellular membrane
   c. sperm tip contacts the egg plasma membrane
   d. decrease in egg's oxygen consumption
   e. increase in the sperm's oxygen consumption

8. Which is a consequence of egg activation?
   a. discarding of a homologous chromosome set in a polar body
   b. formation of an intracellular membrane
   c. sperm flagellum contacts egg plasma membrane
   d. increase in egg's oxygen consumption
   e. decrease in sperm's oxygen consumption

9. How does the size of a 16-celled sea urchin embryo compare to that of the zygote?
   a. It is approximately 16 times the zygote's size.
   b. It is approximately 8 times the zygote's size.
   c. It is approximately the same size as the zygote.
   d. It is approximately 1/2 the zygote's size.
   e. It is approximately 1/16 the zygote's size.

10. How does the size of a 16-celled frog embryo compare to that of the zygote?
    a. It is approximately 16 times the zygote's size.
    b. It is approximately 8 times the zygote's size.
    c. It is approximately the same size as the zygote.
    d. It is approximately 1/2 the zygote's size.
    e. It is approximately 1/16 the zygote's size.

11. If the amount of yolk is the primary determinant of blastocoel size, which animal has the smallest blastocoel?
    a. sea urchin
    b. human
    c. frog
    d. whale
    e. snake

12. If the amount of yolk is the primary determinant of blastocoel size, which animal has the largest blastocoel?
    a. lizard
    b. salamander
    c. quail
    d. horse
    e. hawk

13. Your text describes experiments which inhibited transcription during embryo development. What conclusions were drawn from these?
    a. The gray crescent is necessary for normal development.
    b. Development to the blastula stage only uses protein or mRNA already in the cytoplasm of the egg.
    c. Development to the gastrula stage only uses protein or mRNA already in the cytoplasm of the egg.
    d. Transcription begins with the appearance of blastomeres.
    e. Transcription begins immediately after the gray crescent's appearance in the zygote.

14. Your text describes experiments which inhibited transcription during embryo development. What conclusions were drawn from these?
    a. The gray crescent is not necessary for normal development.
    b. Development to the blastula stage only uses protein or mRNA already in the cytoplasm of the sperm.
    c. Development to the gastrula stage only uses protein or mRNA already in the cytoplasm of the sperm.
    d. Transcription begins before the gastrula stage.
    e. Transcription begins immediately after the gray crescent's disappearance from the zygote.

15. Frog gastrulation includes
    a. disappearance of the blastopore.
    b. appearance of blastomeres.
    c. appearance of a groove on the embryo's dorsal side.
    d. formation of a hollow sphere of cells.
    e. disappearance of the archenteron.

16.   Frog gastrulation includes
    a.   appearance of the blastopore.
    b.   disappearance of blastomeres.
    c.   disappearance of a groove on the embryo's dorsal side.
    d.   formation of a solid sphere of cells.
    e.   appearance of the archenteron.

17.   Which is formed from endoderm?
    a.   nervous system
    b.   skeleton
    c.   middle ear
    d.   thyroid gland
    e.   muscles

18.   Which is formed from ectoderm?
    a.   nervous system
    b.   skeleton
    c.   middle ear
    d.   thyroid gland
    e.   muscles

19.   Which appears first during human embryo development?
    a.   blastocyst
    b.   amniotic cavity
    c.   chorion
    d.   chorionic villi
    e.   placenta

20.   Which appears last during human embryo development?
    a.   blastocyst
    b.   amniotic cavity
    c.   chorion
    d.   chorionic villi
    e.   placenta

Concept Map Construction

Construct a concept map for each group of terms. Be sure to include appropriate connector phrases. You may add other terms as necessary and use terms in the plural or singular form.

    1.   mitosis, transcription, blastomere, blastopore, mesoderm
    2.   gray crescent, dorsal lip, ventral ectoderm, induction, neurulation
    3.   nervous system, ectoderm, embryonic disk, chorion, yolk sac

Chapter 33

Mechanisms of Evolution

Section Concept Map

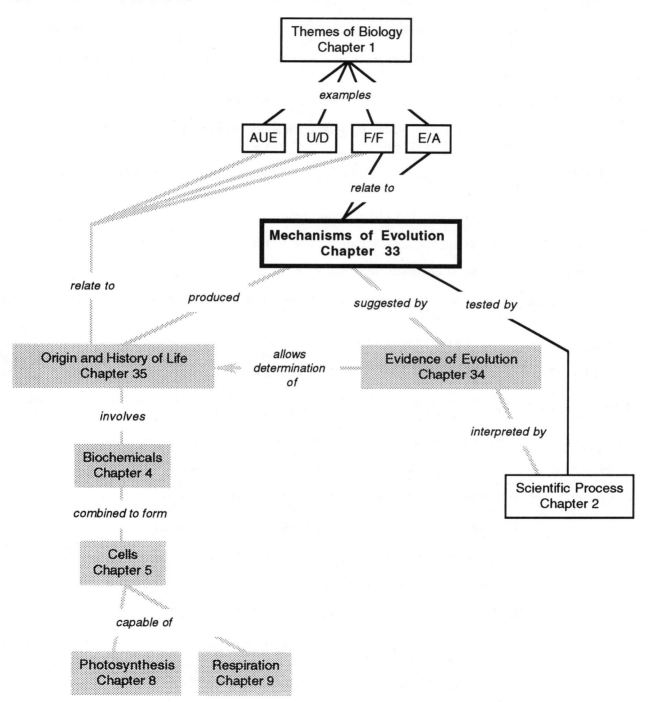

Go Figure!

Figure 33-1: The evidence supplied about the mudskipper is used to support evolution from an aquatic ancestor. Can you supply alternative explanations that are also consistent with the evidence?

Why do you think biologists favor the evolutionary model?

Figure 33-2: What is the probability that a Northern Indian will be homozygous for the I^B allele?

What is the probability that a Native American will be homozygous for the I^B allele?

From the information given, is it possible to calculate frequencies for the other ABO blood group alleles?

Figure 33-5: Which kind of selection is illustrated by the peppered moth population?

Why hadn't the black phenotype disappeared completely prior to the Industrial Revolution in England?

Figure 33-6: How can selection favoring SA heterozygotes keep the frequency of the S allele high in African populations?

Which type of selection does this illustrate?

Figure 33-7: Which type(s) of selection might lead to formation of new species?

Figure 33-8: Under what environmental conditions have these "living fossils" lived?

Figure 33-10: Why are displays such as the ones shown in the figure, or others, such as courtship rituals, characteristic of males?

Figure 33-11: Which human characteristics are examples of sexual dimorphism?

Figure 33-12: Why do zoos send animals on "breeding loan?"

Figure 33-13: Why do you suppose African and Indian elephants are placed in different species (different genera, even!) and these sea stars are not?

What is the main criterion for labeling populations as different species?

Figure 33-14: Under what conditions may divergent evolution occur?

Which type of selection is likely to be involved in divergent evolution?

Besides selection, which other mechanisms may be involved?

Under what conditions may adaptive radiation occur?

Which type of selection is likely to be involved in adaptive radiation?

Under what conditions may convergent evolution occur?

How may a graph of selective pressure (like the ones in Figure 33-7) be modified to illustrate convergent evolution?

Under what conditions may parallel evolution occur?

Which type of selection is likely to be involved for each species?

How may a graph of selective pressure (like the ones in Figure 33-7) be modified to illustrate parallel evolution?

Figure 33-15: Animals as diverse as giraffes, mice, bats and gorillas are classified as members of the same class (Mammalia). Why aren't dolphins, seals, penguins and ichthyosaurs at least placed in the same class since they look more alike than giraffes, mice, bats and gorillas?

Figure 33-17: What does the horizontal axis of these graphs represent?

What do the horizontal branch points indicate about the punctuated equilibrium model?

Why does the gradualism model have acute angles at the branch points?

Matching Questions

Write the letter of the phrase that best matches the numbered term on the left. Use each only once.

_____ 1. evolution

_____ 2. population

_____ 3. gene pool

_____ 4. gene flow

_____ 5. genetic drift

_____ 6. mutation

_____ 7. environment

_____ 8. stabilizing selection

_____ 9. directional selection

_____10. disruptive selection

_____11. sexual selection

_____12. allopatric speciation

_____13. sympatric speciation

_____14. postzygotic isolating mechanism

_____15. reproductive isolation

_____16. divergent evolution

_____17. coevolution

_____18. extinction

_____19. gradualism

_____20. punctuated equilibrium

a. most often seen in an unchanging environment

b. all of the alleles in a population

c. an evolutionary model in which changes occur in small steps

d. random changes in allele frequency

e. an essential step in speciation because it prevents exchange of alleles between populations

f. production of new species in different geographical regions

g. hybrid inviability is an example

h. a form of nonrandom mating

i. physical, physiological or behavioral change in a population over time

j. intermediate phenotypes are at a disadvantage

k. pattern of evolution in which two populations act as selective agents on each other

l. the biological unit that evolves

m. polyploidy is an example

n. the agent of natural selection

o. immigration and emigration are examples

p. pattern of evolution in which two species are formed from a single ancestral type

q. one extreme phenotype is favored, but the other is not

r. loss of a species

s. a model of evolution in which speciation occurs in relatively short periods of time

t. the mechanism by which new alleles arise

Multiple Choice Questions

1. All of the alleles in a population are called the
    a. gene pool.
    b. genetic drift.
    c. allele frequency.
    d. species.
    e. polypeptides.

2. How common an allele is in a population is called the
    a. gene pool.
    b. genetic variation.
    c. allele frequency.
    d. dominant allele.
    e. recessive allele.

NOTE: Answers for questions 3 and 4 are provided in the Appendix. Questions 5 and 6 constitute a similar sequence, but the answers are not given.

3. A certain gene has two alleles. Twenty-five out of every 100 individuals are heterozygous for this gene, whereas 5 out of every 100 are homozygous recessive. What is the frequency of the recessive allele in a population of 100,000 individuals?
    a. 3%
    b. 30%
    c. 35%
    d. 17.5%
    e. 1.75%

4. What is the frequency of the dominant allele in question #3?
    a. 82.5%
    b. 98.25%
    c. 97%
    d. 70%
    e. 65%

5. There is another gene in the population that has two alleles. Twenty-five out of every 75 individuals are heterozygous for this gene, whereas 15 out of every 150 individuals are homozygous recessive. What is the frequency of the recessive allele in a population of 1000 individuals?
    a. 10%
    b. 26.5%
    c. 33.3%
    d. 40%
    e. 50%

6. What is the frequency of the dominant allele in question #5?
    a. 50%
    b. 60%
    c. 66.7%
    d. 73.5%
    e. 90%

7. Which is a requirement for a population to reach genetic (Hardy-Weinberg) equilibrium?
   a. The population must be very large.
   b. The population must not mate.
   c. All immigration into the population must be balanced by emigration out of the population.
   d. Stabilizing selection is the only kind of selection that can occur.
   e. All mutations must produce beneficial alleles.

8. Which is a requirement for a population to reach genetic (Hardy-Weinberg) equilibrium?
   a. Sexual selection is the only kind of selection that can occur.
   b. Diversifying selection is the only kind of selection that can occur.
   c. Directional selection is the only kind of selection that can occur.
   d. The population may be small.
   e. Mutation must not occur.

9. Which is an example of gene flow?
   a. migration
   b. nonrandom mating
   c. sexual selection
   d. chance fluctuations in allele frequencies
   e. polyploidy

10. Which is an example of genetic drift?
    a. migration
    b. nonrandom mating
    c. sexual selection
    d. chance fluctuations in allele frequencies
    e. polyploidy

11. Which describes stabilizing selection?
    a. It will result in speciation if given enough time.
    b. Extreme phenotypes are at an advantage.
    c. The least common phenotype is at an advantage.
    d. It is most often observed in an unchanging environment.
    e. The most common phenotype is at a disadvantage.

12. Which describes disruptive selection?
    a. It will not produce change in the population.
    b. It is common in populations with a large geographic range.
    c. The average phenotype is at an advantage.
    d. Both extreme phenotypes are at a disadvantage.
    e. Change in the population occurs in a single direction.

13. *Festuca viridula* and *Festuca reflexa* belong to the grass family (Gramineae). *F.viridula* is found in subalpine meadows above 6000 feet and flowers from July to August. *F. reflexa* grows in dry, open places below 5000 feet, and flowers from April to June. Based upon the information given, which describes these fescues?
    a. They are reproductively isolated by a prezygotic isolating mechanism.
    b. They are not geographically isolated.
    c. They are the products of stabilizing selection.
    d. They are the products of convergent evolution.
    e. They are the products of allopatric speciation.

14. *Lupinus Tidestromii* and *Lupinus gracilentus* belong to the legume family of angiosperms. *L. Tidestromii* is found in the coastal strand habitat and flowers from May to June, whereas *L. gracilentus* is found in subalpine habitats above 8000 feet and flowers from July to August. Based upon the information given, which describes these lupines?
    a. They are reproductively isolated by postzygotic isolating mechanism.
    b. They are geographically isolated.
    c. They are the products of convergent evolution.
    d. They are the products of sympatric speciation.
    e. They are the products of parapatric speciation.

15. The species of bromegrass (genus *Bromus*) exhibit a variety of chromosome numbers. *B. ciliatus, B. pseudolaevipes, B. vulgaris* (2n = 14), *B. mollis, B. Richardsonii, B. mollis* (2n = 28), *B. catharticus, B. arizonicus, B. marginatus* (2n = 42), and *B. inermis, B. breviaristatus, B. carinatus, B. erectus* (2n = 56) are several examples. Which is consistent with the information given?
    a. The bromegrasses provide an example of polymorphism.
    b. The bromegrasses provide an example of polyploidy.
    c. The populations had to be geographically isolated prior to speciation.
    d. The number of chromosomes in the common ancestor was probably 2n = 7.
    e. The number of chromosomes in the common ancestor was probably 2n = 28.

16. Bluegrasses belong to the genus *Poa*. Species of *Poa* have a variety of chromosome numbers, including *P. annua, P. Bolanderi, P. rhizomata* (2n = 28), *P. confinis, P. compressa, P. Hanseni* (2n = 42), *P. Kelloggii* (2n = 56), and *P. epilis* (2n = 56 or 84). Which is consistent with the information given?
    a. They provide an example of polymorphism.
    b. They provide an example of convergence.
    c. The number of chromosomes in the common ancestor was probably 2n = 14.
    d. The number of chromosomes in the common ancestor was probably 2n = 28.
    e. They provide an example of genetic drift.

Use the phylogenetic tree below for questions 17 and 18.

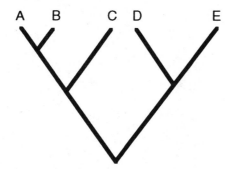

17. Which is consistent with the phylogenetic tree?
    a. Species A and B are the most closely related.
    b. Species C and D are the most closely related.
    c. Species D is more closely related to Species C than to Species E.
    d. If Species A and C belong to one family, then Species B belongs to another family.
    e. Species A and B are the products of directional selection.

18. Which is consistent with the phylogenetic tree?
    a. Species A is more closely related to Species C than to Species B.
    b. If Species B and D belong to one genus, then Species C belongs to another genus.
    c. Species A, B and C are the products of adaptive radiation.
    d. Species D and E are the products of convergent evolution.
    e. Species D and E are the products of disruptive selection.

19. The theory of punctuated equilibrium is best supported by a fossil record with
    a. morphological forms disappearing, then reappearing.
    b. long periods where the fossils show little change separated by short periods where there is a lot of change.
    c. numerous intermediate forms which suggest a gradual change.
    d. numerous different morphological forms in each period.
    e. very little change in the fossils.

20. Phyletic evolution is best supported by a fossil record with
    a. numerous gaps between morphological forms.
    b. long periods where the fossils show little change separated by short periods where there is a lot of change.
    c. numerous intermediate forms which suggest a gradual change.
    d. numerous different morphological forms in each period.
    e. very little change in the fossils.

Concept Map Construction

Construct a concept map for each group of terms. Be sure to include appropriate connector phrases. You may add other terms as necessary and use terms in the singular or plural form.

1. diversifying selection, population, gene flow, Hardy-Weinberg Equilibrium, allele frequency
2. sympatric speciation, reproductive isolation, gene pool, parallel evolution, dominant allele
3. punctuated equilibrium, extinction, sexual selection, meiosis, DNA

Chapter 34

Evidence For Evolution

Section Concept Map

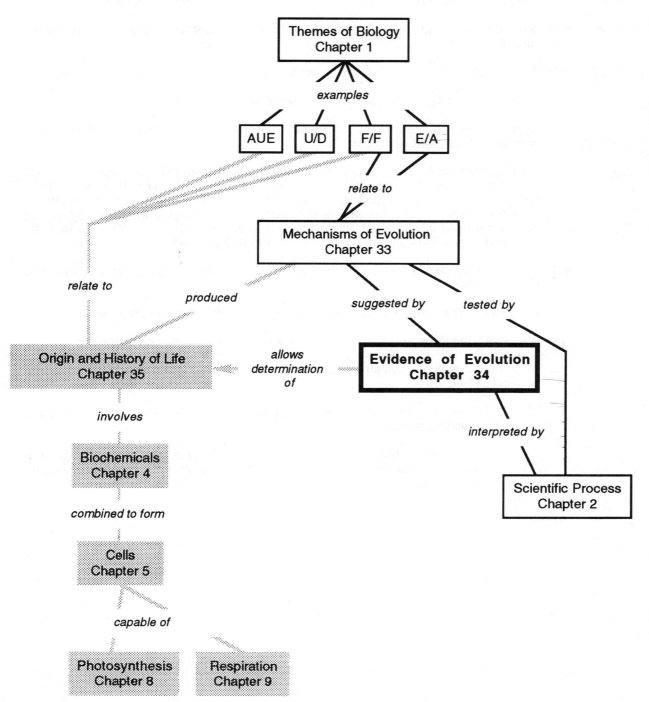

Go Figure!

Figure 34-1: Since the cycad and the true palm are not closely related, are their palm-like characteristics homologous or analogous?

Are the leaves themselves homologous or analogous?

Figure 34-2: Were the species of artiodactyla produced by divergent evolution, convergent evolution, or adaptive radiation?

Figure 34-3: Why do you suppose the forelimb skeletons are not compared to a primitive vertebrate such as a fish?

Figure 34-5: What reptilian features did *Archaeopterix* have?

What avian (bird) features did *Archaeopterix* have?

If only skeletal remains had been found, how would *Archaeopterix* probably have been classified - bird or reptile?

Figure 34-6: Jean Baptiste de Lamarck proposed a theory of evolution prior to Charles Darwin. The theory involved inheritance of acquired characteristics. That is, modifications accumulated by an organism throughout its life could be passed on to its offspring. Lamarck might modify this proposal to account for vestigial structures by saying organs not used by an organism during its life are passed on to offspring in some atrophied form. How might Darwin respond to this proposal?

Figure 34-8: What are the major differences between the skulls of *Australopithecus africanus* and *Homo sapiens*?

Which skull does *Homo erectus* most closely resemble? Defend your answer.

Figure 34-11: Why don't the shadows of ancestral species overlap with the descendants?

Why are the australopithecines represented across the diagram in the 2 - 4 million year range?

Why is *Homo erectus* at the same end of the diagram as *Homo habilis* and not nearer the australopithecines?

(Just a note: modern humans arose between 200,000 and 100,000 years ago, but they have been involved in agriculture - a prerequisite to complex civilization - only about 10,000 years.)

Matching Questions

Write the letter of the phrase that best matches the numbered term on the left. Use each only once.

_____ 1. hominid

_____ 2. homologous feature

_____ 3. homoplastic (analogous) feature

_____ 4. primitive feature

_____ 5. derived feature

_____ 6. fossil record

_____ 7. law of stratigraphy

_____ 8. biogeography

_____ 9. vestigial structure

_____10. *Archaeopterix*

_____11. australopithecines

_____12. Piltdown Man

_____13. *Homo habilis*

_____14. *Australopithecus afarensis*

_____15. *Homo erectus*

_____16. *Homo sapiens*

a. similarity of two species because of convergent evolution

b. a reduced organ that has no apparent function in an organism

c. the study of distribution of organisms

d. a group of early hominids considered by many not to be on the lineage leading to modern humans

e. a hominid species that lived 3.5 million years ago

f. general term applied to humans and their recent ancestors that walked upright

g. older rocks are nearer the bottom of an undisturbed geologic formation

h. modern humans

i. the arrangement seen in the ancestor

j. remains used by paleontologists to piece together evolutionary history

k. similarity of two species because of a common ancestor

l. modified from the ancestral arrangement

m. a fossil forgery

n. a fossil bird with many reptilian characteristics

o. Peking Man and Java Man belong to this species

p. the earliest known hominid species to use tools

Multiple Choice Questions

1. Structures which have a similar construction and embryonic origin, but not necessarily a similar function are said to be
   a. homoplastic.
   b. analogous.
   c. deleterious.
   d. homologous.
   e. analytical.

2. Structures that look alike due to a similar function but not because of common ancestry are said to be
   a. homoplastic.
   b. homologous.
   c. homosporous.
   d. analgesic.
   e. anagogic.

3. Lions and tigers belong to the same mammalian order (Carnivora) and animal subphylum (Vertebrata). Which is most likely true of lions and tigers?
   a. The cat-like similarities are due to convergent evolution.
   b. The differences are due to divergent evolution.
   c. They are more closely related to elephants (order Proboscidea) than to each other.
   d. Their vertebral columns are analogous structures.
   e. They share a common ancestor with stripes and a mane.

4. Whales and dolphins (small whales) belong to the same order of aquatic mammals (Cetacea). Which is most likely true of whales and dolphins?
   a. The similarities of a fish-like body are due to convergent evolution.
   b. The differences are due to divergent evolution.
   c. They are more closely related to seals (order Pinnipedia) than to each other.
   d. Their lungs are analogous structures.
   e. They do not share a common ancestor with fish-like body features.

5. Many members of the plant family Euphorbiaceae are cactus-like in appearance. True cacti belong to the family Cactaceae. Both are angiosperm families. Which most likely describes euphorbs and cacti?
   a. The spines are the result of convergent evolution.
   b. The presence of flowers in both is the result of convergent evolution.
   c. The presence of xylem and phloem is the result of convergent evolution.
   d. The stamens in both are analogous structures.
   e. The roots are analogous structures.

6. Penguins are birds (class Aves) and seals are mammals (class Mammalia). Both belong to the phylum Chordata and have a vertebral column surrounding a spinal cord. Which is true of penguins and seals?
   a. The fins are the product of convergent evolution.
   b. The presence of the vertebral column is the result of convergent evolution.
   c. The flippers are homologous structures.
   d. The streamlined body shapes are homologous structures.
   e. The spinal cords are analogous structures.

7. Sharks are fish (class Chondryichthyes) and have gills, whereas dolphins are mammals (class Mammalia) and have lungs. Both belong to the phylum Chordata and have a vertebral column. Which is most likely true of sharks and dolphins?
    a. The vertebral columns are analogous structures.
    b. The dorsal fins are homologous structures.
    c. The tail fins are the product of convergent evolution.
    d. Lungs and gills are analogous structures.
    e. Lungs and gills are homologous structures.

8. Whales belong to the mammalian order Cetacea, while seals belong to the mammalian order Pinnipedia. The evidence indicates that both evolved from different terrestrial mammalian ancestors. Which is most likely true of whales and seals?
    a. The front fins are homologous structures.
    b. The forelimb bones are homologous structures.
    c. The presence of hair follicles is the result of convergent evolution.
    d. The ability to maintain a constant body temperature is the result of convergent evolution.
    e. The ability to breathe air is the result of convergent evolution.

9. Compared to other mammals, which describes seals (class Mammalia, order Pinnipedia)?
    a. Fur is a derived characteristic.
    b. Presence of lungs is a derived characteristic.
    c. The vertebral column is a primitive characteristic.
    d. Presence of flippers is a primitive characteristic.
    e. The fish-like tail is a primitive characteristic.

10. Compared to other flowering plants, which describes cacti (division Anthophyta, family Cactaceae)?
    a. Flowers are derived structures.
    b. Xylem and phloem are primitive structures.
    c. Roots are derived structures.
    d. Chloroplasts are derived structures.
    e. Spines (modified leaves) are primitive structures.

11. The law of stratigraphy provides a means of
    a. determining absolute age of fossils.
    b. determining relative age of fossils.
    c. distinguishing between homologous structures.
    d. distinguishing between analogous structures.
    e. distinguishing between birds and reptiles.

12. Radioisotopes provide a means of
    a. determining absolute age of fossils.
    b. determining relative age of fossils.
    c. distinguishing between homologous structures.
    d. distinguishing between analogous structures.
    e. distinguishing between birds and reptiles.

13. Which is most likely to become fossilized?
    a. herbaceous plant
    b. flower
    c. tooth
    d. jelly fish
    e. earthworm

14. Which is most likely to become fossilized?
    a. a desert insect
    b. a marine clam
    c. a flowering plant from the mountains
    d. prairie grass
    e. a desert snake

15. According to current evidence, which is the oldest species?
    a. *Homo habilis*
    b. *Homo erectus*
    c. *Homo sapiens*
    d. *Australopithecus robustus*
    e. *Australopithecus afarensis*

16. According to current evidence, which is the youngest species?
    a. *Homo habilis*
    b. *Homo erectus*
    c. *Homo sapiens*
    d. *Australopithecus robustus*
    e. *Australopithecus afarensis*

Concept Map Construction

Construct a concept map for each group of terms. Be sure to include appropriate connector phrases. You may add other terms as necessary and use terms in the singular or plural form.

1. homology, convergent evolution, vestigial structure, comparative embryology, biogeography
2. *Homo habilis*, gene pool, natural selection, fossil record, adaptive radiation
3. reproductive isolation, bone, radioactive carbon, law of stratigraphy, *Australopithecus afarensis*

Chapter 35

The Origin and History of Life

Section Concept Map

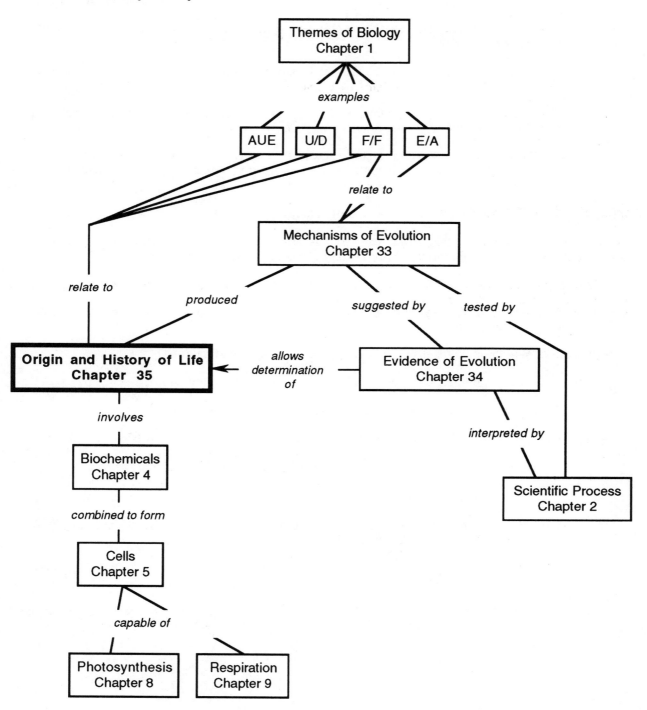

Go Figure!

Figure 35-1: What is the glowing sphere in the center?

What are the smaller swirls?

What do the rings represent?

Figure 35-3: How much longer was the Proterozoic Era than the other three eras combined?

Why do geologists cram three eras in a time span that is only a fraction as long as the Proterozoic?

What characteristics of Proterozoic organisms might make it difficult for geologists to break this era into smaller time spans?

What correlation between existing species diversity, competition intensity, and evolutionary rate is suggested by this chart?

Figure 35-4: Why do the authors say that "it has been proposed that a colonial organism such as *Volvox* gave rise to all multicellular organisms" rather than "it has been proposed that *Volvox* gave rise to all multicellular organisms?"

Figure 35-5: Why are fossils of soft-bodied organisms relatively rare?

The age of this fossil is given as approximately 650 million years. How are biologists able to give estimates of fossil age in years?

How many years represents a 1% error in the estimate of this fossil's age?

Figure 35-6: Where is older sea floor located: at the mid-Atlantic ridge or near the Atlantic coast of North America?

Figure 35-7: Which geologic structure is associated with a spreading zone?

Which geologic structure is associated with a subduction zone?

What are three possible consequences of continental collision?

Figure 35-8: For at least how many years has North America been separated from Africa?

At what point did India separate from Gondwanaland?

Was the joining of India and Asia a recent geologic event?

Figure 35-9: Which vertebrate groups (fish, amphibians, reptiles, birds, or mammals) first appeared during the Paleozoic Era?

Which major plant groups (nonseed plants, gymnosperms, and flowering plants) first appeared during the Paleozoic era?

Assume the first vascular plants appeared at the beginning of the Silurian Period. What percent of earth's existence have vascular plants been living?

Figure 35-10: Assume mammals first appeared at the beginning of the Triassic Period. What percent of earth's existence have mammals been living?

Dinosaurs were dominant land animals for 125 million years. What percent of earth's existence does this represent?

Figure 35-12: In Chapter 34 we saw that *Australopithecus afarensis* ("Lucy") lived 3.5 million years ago. During which geologic period was this?

Assume modern humans (*Homo sapiens*) first appeared on earth 200,000 years ago (currently the earliest estimate of our arrival). Life first appeared 3.5 billion years ago. What percentage of that time have modern humans been living?

Matching Questions

Write the letter of the phrase that best matches the numbered term on the left. Use each only once.

_____ 1.   chemical evolution

_____ 2.   3.5 billion years

_____ 3.   prokaryotic cells

_____ 4.   mass extinction

_____ 5.   cyanobacteria

_____ 6.   plate tectonics

_____ 7.   Pangaea

_____ 8.   subduction

_____ 9.   Mesozoic Era

_____10.   Proterozoic Era

_____11.   Cenozoic Era

_____12.   Pleistocene Epoch

_____13.   pterosaur

_____14.   therapsid

_____15.   stromatolite

a.   age of the oldest fossils

b.   reptile group thought to be ancestral to the mammals

c.   group responsible for producing the aerobic atmosphere

d.   characterized by periods of glaciation

e.   first flying vertebrate

f.   reptiles dominated this time in history

g.   a mass of prokaryotic cells and minerals existing today, but also found as fossil evidence of early life

h.   mammals dominate this time in history

i.   the formation of life from the development of increasingly complex molecules

j.   these gave rise to eukaryotic cells 2 billion years ago

k.   process of one of earth's plates sliding under another

l.   a period of time in which many groups of organisms die out

m.   geologic time period in which the first cells evolved

n.   a model to explain continental drift

o.   a supercontinent that existed 200 million years ago

Multiple Choice Questions

1. Which DOES NOT describe the formation of the sun?
    a. It formed from matter.
    b. The gravitational attraction between particles caused it to form.
    c. Thermonuclear reactions occurred as a result of particles coming together.
    d. The sun formed 3.5 million years ago.
    e. The sun formed at the center of the solar system.

2. Which DOES NOT describe the formation of the earth?
    a. Its atmosphere vaporized due to intense heat from the sun.
    b. It condensed from the same matter that produced the sun.
    c. The earth was initially very hot.
    d. It formed at the same time as the sun.
    e. It was in an ideal position relative to the sun to support life.

3. Which DOES NOT describe the Proterozoic Era?
    a. The first multicellular organisms were formed.
    b. The first cells arose.
    c. Oxygen accumulated in the atmosphere.
    d. It is the shortest of the four geologic eras.
    e. The first prokaryotes arose.

4. Which DOES NOT describe the Proterozoic Era?
    a. Competition for resources and natural selection did not occur.
    b. The atmosphere initially lacked oxygen.
    c. Chemical evolution occurred.
    d. It covers a four billion year time period beginning with the origin of the earth.
    e. The first eukaryotes arose.

5. The first heterotrophs lived during the
    a. Proterozoic Era.
    b. Paleozoic Era.
    c. Mesozoic Era.
    d. Cambrian Period.
    e. Jurassic Period.

6. The first autotrophs lived during the
    a. Pleistocene Epoch.
    b. Triassic Period.
    c. Proterozoic Era.
    d. Devonian Period.
    e. Permian Period.

7. Of these five, which came first?
    a. reptiles
    b. mammals
    c. autotrophic prokaryotes
    d. multicellular eukaryotes
    e. heterotrophic prokaryotes

8. Of these five, which came first?
    a. anaerobic respiration
    b. aerobic respiration
    c. anaerobic photosynthesis
    d. aerobic photosynthesis
    e. fermentation

9.   Which DOES NOT describe multicellularity?
     a.   Evidence suggests that it developed only once.
     b.   It is found only in eukaryotic organisms.
     c.   It enables cells to perform specialized functions.
     d.   Some multicellular organisms arose from colonies of cells.
     e.   Multicellular organisms first appeared in the Proterozoic Era.

10. Which DOES NOT describe multicellularity?
     a.   Some multicellular organisms arose from aggregations of independent cells.
     b.   Multicellular organisms arose about one billion years ago.
     c.   Multicellular prokaryotes gave rise to multicellular eukaryotes in the late Proterozoic Era.
     d.   Multicellularity preceded the development of the major animal phyla.
     e.   Multicellularity preceded the development of the major plant phyla.

11. Which IS NOT consistent with the theory of continental drift?
     a.   It can account for unusual fossil distributions.
     b.   Plate tectonics provides a possible mechanism.
     c.   It was first proposed in the early 20th century.
     d.   Pangaea was a supercontinent that existed about 200 million years ago.
     e.   Pangaea split into Laurasia and Magma about 150 million years ago.

12. Which IS NOT consistent with the theory of plate tectonics?
     a.   The edge of some continental plates slide under others at subduction zones.
     b.   The earth's crust is fragmented into relatively thick continental plates.
     c.   Formation of new ocean floor pushes plates apart.
     d.   Mountains may be built when plates collide.
     e.   Plate movement may account for earthquakes.

13. Of these five, which came first?
     a.   reptiles
     b.   fish
     c.   amphibians
     d.   birds
     e.   mammals

14. Of these five, which came first?
     a.   algae
     b.   land plants
     c.   vascular plants
     d.   seed plants
     e.   flowering plants

15. The first land plants lived during the
     a.   Paleozoic Era
     b.   Jurassic Period
     c.   Cambrian Period
     d.   Silurian Period
     e.   Triassic Period

16. The first vertebrates lived during the
     a.   Mesozoic Era
     b.   Permian Period
     c.   Ordovician Period
     d.   Devonian Period
     e.   Pleistocene Epoch

17. Dinosaurs were the dominant land animals during the
    a. Jurassic Period
    b. Triassic Period
    c. Cretaceous Period
    d. Tertiary Period
    e. Quaternary Period

18. Mammals became the dominant land mammals during the
    a. Jurassic Period
    b. Triassic Period
    c. Cretaceous Period
    d. Tertiary Period
    e. Quaternary Period

19. Gymnosperms were the dominant land plants during the
    a. Cambrian Period
    b. Triassic Period
    c. Cretaceous Period
    d. Ordovician Period
    e. Permian Period

20. Flowering plants first became abundant during the
    a. Cambrian Period
    b. Triassic Period
    c. Cretaceous Period
    d. Ordovician Period
    e. Permian Period

Concept Map Construction

Construct a concept map for each group of terms. Be sure to include appropriate connector phrases. You may add other terms as necessary and use terms in the singular or plural form.

1. divergent evolution, chemical evolution, geologic time scale, origin of life, prokaryote
2. *Homo sapiens*, continental drift, subduction zone, therapsid, adaptive radiation
3. autotroph, homeostasis, dinosaur, mass extinction, natural selection

Chapter 36

The Monera Kingdom and Viruses

Section Concept Map

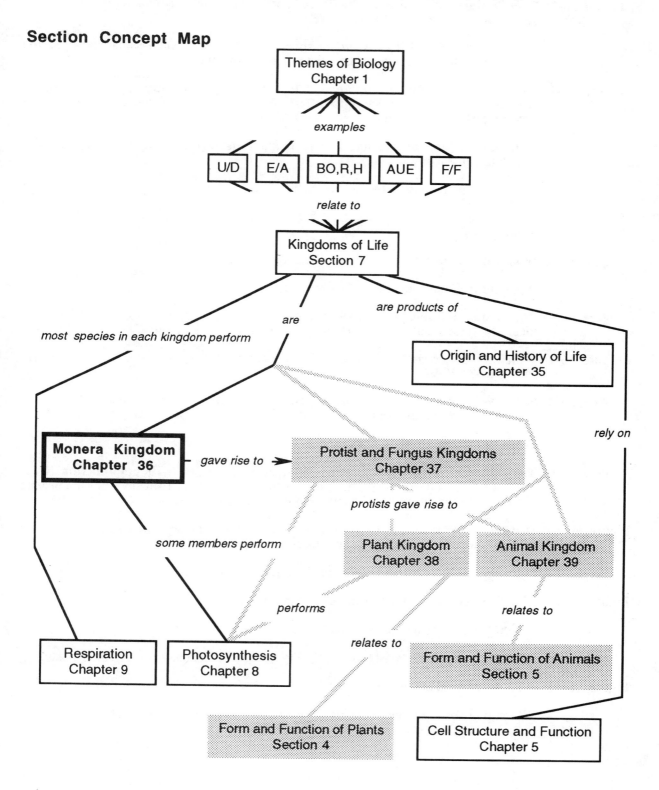

Go Figure!

Figure 36-2: The bacteria in Figure 36-2a form clusters. Why are these bacteria not considered by biologists to be multicellular organisms?

Figure 36-3: Why do the authors suggest that these modern anaerobic species of bacteria might be similar to the earliest forms of life?

What does the fact that *Clostridium* species live in places such as canned food and wounded flesh suggest about their type of metabolism?

Figure 36-4: The caption title for this figure is "One proposal of evolutionary descent of all organisms from early prokaryotic cells." What does that tell you about the current state of classification at the kingdom level?

Is it possible to have different descent patterns where each is consistent with available evidence?

Which groups belong to the eubacteria?

What evidence supports the claim that archaebacteria are not descendents of other prokaryotes?

Figure 36-5: What cell shapes are shown in these photomicrographs?

What are the main functions of bacterial stains?

Figure 36-6: Why should the results of biochemical tests run on a single colony of bacteria be identical to the results run on any single cell in the same colony?

Figure 36-7: Which structures shown do you consider essential for a bacterium to be minimally functional? Why?

Figure 36-8: Prokaryotes lack membranous organelles. Why are the stacks of internal membranes in this cyanobacterium not considered to be membranous organelles?

What membrane are they made of?

Figure 36-9: The nitrogen-fixing enzyme is sensitive to oxygen and will not function in its presence. Which part(s) of the photosynthetic apparatus in a heterocyst must be disabled to prevent oxygen inactivation of the enzyme?

Figure 36-10: Is it likely the magnetic crystals of homing pigeons, honeybees and the bacterium are homologous or analogous structures?

Is this an example of convergent or divergent evolution?

Figure 36-12: Which structures are present in viruses and bacteria?

Which bacterial structures are absent from viruses?

What is the molecular basis for specificity between virus particles and their hosts?

Figure 36-13: What is the viral eclipse phase?

Which parts of the virus replicative cycle belong to the eclipse phase?

Matching Questions

Write the letter of the phrase that best matches the numbered term on the left. Use each only once.

_____ 1.   eubacteria

_____ 2.   nucleoid

_____ 3.   peptidoglycan

_____ 4.   coccus

_____ 5.   prokaryotic fission

_____ 6.   facultative anaerobe

_____ 7.   bacterial chlorophyll

_____ 8.   cyanobacteria

_____ 9.   heterocyst

_____10.   methanogens

_____11.   saprophyte

_____12.   virus

_____13.   capsid

_____14.   eclipse phase

_____15.   retrovirus

_____16.   viroid

_____17.   prion

_____18.   vaccine

a. a spherical cell

b. a structure used for nitrogen fixation

c. an organism capable of fermentation or respiration, depending on the availability of oxygen

d. time during the virus replicative cycle in which no complete virus particles are seen

e. infectious RNA with no protein covering

f. infectious protein particles lacking any nucleic acid

g. the pigment of photosynthetic bacteria

h. the group of true bacteria

i. virus capable of making DNA from RNA

j. cell wall material of most prokaryotes

k. these prokaryotes produce oxygen during photosynthesis

l. a general term for a noncellular parasite

m. type of asexual reproduction

n. DNA region of a prokaryotic cell

o. archaebacteria that produce methane gas

p. an organism that gets nutrients from dead matter

q. a protective agent used to stimulate the immune system prior to infection by a pathogenic virus

r. the part of a virus that contains the hereditary material

Multiple Choice Questions

1. Which IS NOT a characteristic of the Kingdom Monera?
   a. peptidoglycan cell wall
   b. prokaryotic ribosomes
   c. rapid growth rate
   d. a variety of metabolic capabilities
   e. true nucleus

2. Which IS NOT a characteristic of the Kingdom Monera?
   a. small size
   b. large population size
   c. asexual reproduction by fission
   d. presence of endoplasmic reticulum
   e. a variety of cell shapes

3. Which name is applied to spherical prokaryotic cells?
   a. spirillum
   b. coccus
   c. vibrio
   d. spirochaete
   e. bacillus

4. Which name is applied to rod-shaped prokaryotic cells?
   a. spirillum
   b. coccus
   c. vibrio
   d. spirochaete
   e. bacillus

5. *Escherichia coli* can divide every 20 minutes if grown in optimum conditions. If you start with one cell in optimum conditions at 8 AM and it divides after 20 minutes, how many cells will you have at 1:30 PM that same day? (Assume no cells die and each cell divides on a synchronous 20 minute schedule.)
   a. 17
   b. 34
   c. 512
   d. 65,536
   e. 131,072

6. Suppose an *E. coli* cell is grown in an environment that changes the time between divisions to 30 minutes. If you start with one cell at 8 AM, how many cells will you have at 1:20 PM that same day? (Assume no cells die and each cell divides on a synchronous 30 minute schedule.)
   a. 2048
   b. 1024
   c. 512
   d. 128
   e. 11

7. Which DOES NOT describe the Eubacteria?
   a. It includes the majority of prokaryotic species.
   b. Eubacteria possess a range of oxygen tolerances.
   c. Photosynthetic eubacteria may use bacterial chlorophyll or chlorophyll *a*.
   d. Eubacteria have a protein cell wall.
   e. Some eubacteria form a capsule to retard water loss.

8. Which DOES NOT describe the Eubacteria?
    a. Some produce endospores.
    b. Most eubacteria are pathogenic.
    c. Flagella act as rotors to produce movement.
    d. Eubacteria lack a membrane-bound nucleus.
    e. Cyanobacteria are considered to be eubacteria.

9. Which describes cyanobacteria?
    a. They have bacterial chlorophyll.
    b. All cells are capable of nitrogen fixation.
    c. They lack chloroplasts.
    d. They have a cellulose cell wall.
    e. They are strictly unicellular.

10. Which describes cyanobacteria?
    a. They are classified with the other algae.
    b. Heterocysts are cells specialized for photosynthesis.
    c. They have a protein cell wall.
    d. They always form filaments of cells.
    e. Their photosynthesis produces oxygen.

11. Which describes the Archaebacteria?
    a. Their cell walls are made of peptidoglycan.
    b. They often inhabit extreme environments.
    c. They have heterocysts that allow them to fix CO_2.
    d. They are eukaryotes.
    e. They include the majority of Monerans.

12. Which describes the Archaebacteria?
    a. They are urkaryotes.
    b. They include cyanobacteria.
    c. Some are thermophilic.
    d. They have typical membranes made of phospholipids.
    e. They have a nuclear membrane around DNA.

13. Which DOES NOT describe viruses?
    a. Each virus has both DNA and RNA in it.
    b. Viruses are obligate intracellular parasites.
    c. They exhibit some, but not all, of the characteristics associated with living organisms.
    d. Replication of a single virus may result in production of hundreds of new viruses.
    e. A protein capsid surrounds the nucleic acid portion of each virus.

14. Which DOES NOT describe viruses?
    a. They are larger than archaebacteria.
    b. The replicative cycle includes a period in which no complete virus particles are present.
    c. They infect host cells.
    d. Host cell nucleic acid polymerases are used during viral replication.
    e. They are not cellular organisms.

15. Which disease is caused by rhinoviruses?
    a. common cold
    b. influenza
    c. AIDS
    d. measles
    e. mumps

16. Which disease is caused by herpesviruses?
    a. cold sores
    b. influenza
    c. AIDS
    d. measles
    e. mumps

17. Which is an infectious protein particle?
    a. virion
    b. retrovirus
    c. viroid
    d. prion
    e. capsid

18. Which is an infectious RNA particle?
    a. virion
    b. retrovirus
    c. viroid
    d. prion
    e. capsid

19. What is the most effective protection against viral infections?
    a. disinfectants
    b. natural immunity of the host
    c. antibiotics
    d. analgesics
    e. washing hands

20. Why are most chemical treatments generally ineffective at combating viral infections?
    a. Viruses are not made of chemicals, so chemical treatments are ineffective.
    b. Viruses are not made of the same chemicals as living organisms, so chemical treatments are ineffective.
    c. Viruses enter host cells and it is difficult to attack them without damaging the host.
    d. Viruses have enzymes that break down most antiviral chemicals.
    e. Viruses have a selectively permeable membrane that prevents entry of most antiviral chemicals.

Concept Map Construction

Construct a concept map for each group of terms. Be sure to include appropriate connector phrases. You may add other terms as necessary and use terms in the singular or plural form.

1. peptidoglycan, nucleoid, Monera, eubacteria, DNA
2. prokaryotic fission, facultative anaerobe, chemosynthesis, cyanobacteria, heterocyst
3. capsid, DNA, viroid, prion, eclipse phase

Chapter 37

The Protist and Fungus Kingdoms

Section Concept Map

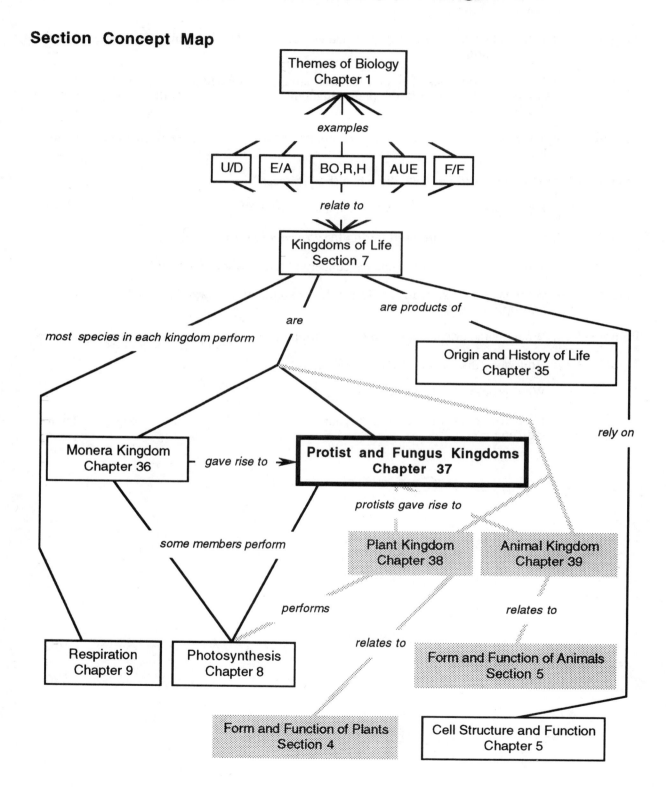

Themes of Biology
Chapter 1

examples

U/D   E/A   BO,R,H   AUE   F/F

relate to

Kingdoms of Life
Section 7

most species in each kingdom perform

are

are products of

Origin and History of Life
Chapter 35

rely on

Monera Kingdom
Chapter 36

gave rise to

**Protist and Fungus Kingdoms
Chapter 37**

protists gave rise to

Plant Kingdom
Chapter 38

Animal Kingdom
Chapter 39

some members perform

performs

relates to

relates to

Respiration
Chapter 9

Photosynthesis
Chapter 8

Form and Function of Animals
Section 5

Form and Function of Plants
Section 4

Cell Structure and Function
Chapter 5

Go Figure!

Figure 37-2: Why is *Euglena* placed in the protist kingdom rather than the plant kingdom?

What characteristics are visible in the picture that eliminate *Euglena* from the kingdom Monera?

Figure 37-3: If these two *Euplotes* simply exchange genetic material, is this an example of sexual reproduction?

Figure 37-4: Your text suggests that multicellular algae are placed in the protist kingdom by some authorities and in the plant kingdom by others. How is this issue addressed in the illustration?

Figure 37-5: To which of the four protozoan groups shown in Figure 37-4 do these organisms belong?

Figure 37-6: Why does a trophozoite die when shed in feces?

How is the cyst able to survive outside the host's body?

Figure 37-7: How many hosts are used during the *Plasmodium* life cycle?

Why are public health agencies concerned with mosquito control?

Figure 37-9: Why is it energetically efficient for an animal as large as a whale to feed on phytoplankton?

Figure 37-11: What characteristic shown in the picture indicates the organism is a dinoflagellate?

Figure 37-18: How does dikaryotic differ from diploid?

What process produces basidiospores?

Is kicking over a mushroom in your lawn an effective way of ridding your lawn of the fungus?

Matching Questions

Write the letter of the phrase that best matches the numbered term on the left. Use each only once.

_____ 1. zooplankton

_____ 2. trophozoite

_____ 3. cyst

_____ 4. pseudopodia

_____ 5. diatoms

_____ 6. dinoflagellates

_____ 7. cellular slime mold

_____ 8. plasmodial slime mold

_____ 9. mycorrhizae

_____10. hypha

_____11. lichen

_____12. zygospore

_____13. ascospore

_____14. basidiospore

_____15. deuteromycete

a. temporary extensions of cytoplasm used for motility in some protozoans

b. algae with silica cell walls

c. produced in mushrooms

d. only diploid cell in the zygomycete life cycle

e. a group of unicellular heterotrophs

f. a filament of fungus cells

g. an association between a fungus and an alga

h. these are characterized by having two flagella

i. fungus in which no evidence of sexual reproduction has been discovered

j. the active form of a polymorphic protozoan

k. one part of its life cycle includes a large, multinucleated "cell"

l. a reproductive cell of ascomycetes

m. the dormant form of a polymorphic protozoan

n. an association between a fungus and a vascular plant

o. life cycle includes an amoeba-like stage and a fruiting stage

Multiple Choice Questions

1. Which is an evolutionary advancement credited to protists?
   a. fermentation
   b. photosynthesis
   c. DNA replication
   d. sexual reproduction
   e. ribosomes

2. Which is an evolutionary advancement credited to protists?
   a. photosynthesis
   b. chloroplasts
   c. cell membrane
   d. cell division
   e. motility

3. Which describes a "typical" protozoan?
   a. multicellular
   b. cellulose cell wall
   c. photosynthetic
   d. ingestive heterotroph
   e. nonmotile

4. Which describes a "typical" protozoan?
   a. peptidoglycan cell wall
   b. chitinous cell wall
   c. autotrophic
   d. pathogenic to humans
   e. motile

5. Which DOES NOT describe a trophozoite?
   a. It is the active, growing part of the life cycle of certain protozoans.
   b. It is typical of polymorphic protozoans.
   c. It will eventually develop into a cyst.
   d. It is the feeding stage of the life cycle of certain protozoans.
   e. It is best adapted to survival in dry environments.

6. Which DOES NOT describe a cyst?
   a. It is killed by severe environmental conditions the trophozoite can withstand.
   b. It is a dormant part of the life cycle of certain protozoans.
   c. It is typical of polymorphic protozoans.
   d. It is surrounded by a thick wall.
   e. It will eventually develop into a trophozoite.

7. Which DOES NOT describe zooplankton?
   a. Most are photosynthetic.
   b. They are important members of aquatic food chains.
   c. They are eukaryotic.
   d. Many are decomposers.
   e. They are unicellular.

8. Which DOES NOT describe phytoplankton?
   a. Most are photosynthetic.
   b. They are important members of aquatic food chains.
   c. They are prokaryotic.
   d. They produce the majority of atmospheric oxygen.
   e. They are microscopic.

9. To which group does a single celled, photosynthetic protist with two flagella belong?
   a. blue-green algae
   b. dinoflagellate
   c. diatom
   d. ciliate
   e. Sarcodina

10. To which group does a single celled, photosynthetic protist with silica cell walls belong?
    a. blue-green algae
    b. dinoflagellate
    c. diatom
    d. flagellate
    e. ascomycete

11. Which DOES NOT describe cellular slime molds?
    a. One stage of the life cycle involves amoeboid cells.
    b. They are heterotrophic.
    c. Individual cells come together to form a fruiting body when environmental conditions are poor.
    d. Cyclic AMP is a chemical used to organize development during one stage of the life cycle.
    e. Spore germination produces a fruiting body.

12. Which DOES NOT describe plasmodial slime molds?
    a. One stage of the life cycle involves amoeboid cells.
    b. They are heterotrophic.
    c. One stage of the life cycle involves a large, multinucleate cell.
    d. The plasmodium is the feeding stage of the life cycle.
    e. The plasmodium develops from a zygote.

13. Which describes fungi?
    a. Most are autotrophic.
    b. Most are motile.
    c. Most are unicellular.
    d. Most are parasitic.
    e. Most are capable of sexual reproduction.

14. Which describes fungi?
    a. Typically, the only diploid cell is the zygote.
    b. Photosynthetic fungi are important members of terrestrial food chains.
    c. Unicellular fungi form hyphae when mature.
    d. Haploid spores combine to form a diploid zygote.
    e. Spores are the only haploid cells in most fungal life cycles.

15. Which terrestrial fungal group is characterized by nonseptate hyphae and the absence of flagella?
    a. ascomycetes
    b. zygomycetes
    c. basidiomycetes
    d. deuteromycetes
    e. oomycetes

16. Which terrestrial fungal group is characterized by sexual spores contained in an ascus?
    a. ascomycetes
    b. zygomycetes
    c. basidiomycetes
    d. deuteromycetes
    e. oomycetes

17.   Which terrestrial fungal group is responsible for producing mushrooms and puffballs?
    a.   ascomycetes
    b.   zygomycetes
    c.   basidiomycetes
    d.   deuteromycetes
    e.   oomycetes

18.   Which terrestrial fungal group has no apparent sexual component in the life cycle?
    a.   ascomycetes
    b.   zygomycetes
    c.   basidiomycetes
    d.   deuteromycetes
    e.   oomycetes

19.   Which describes basidiospores?
    a.   They are diploid and produced by meiosis.
    b.   They are haploid and produced by meiosis.
    c.   They are diploid and produced by mitosis.
    d.   They are haploid and produced by mitosis.
    e.   They are dikaryotic and produced by fertilization.

20.   Which describes ascospores?
    a.   They are diploid and produced by meiosis.
    b.   They are haploid and produced by meiosis.
    c.   They are diploid and produced by mitosis.
    d.   They are haploid and produced by mitosis.
    e.   They are dikaryotic and produced by fertilization.

Concept Map Construction

Construct a concept map for each group of terms. Be sure to include appropriate connector phrases. You may add other terms as necessary and use terms in the singular or plural form.

    1.   prokaryote, cell wall, photosynthesis, phytoplankton, dinoflagellate
    2.   eukaryote, zooplankton, pseudopodium, plasmodium, trophozoite
    3.   heterotroph, lichen, saprobe, basidiospore, deuteromycete

Chapter 38

The Plant Kingdom

Section Concept Map

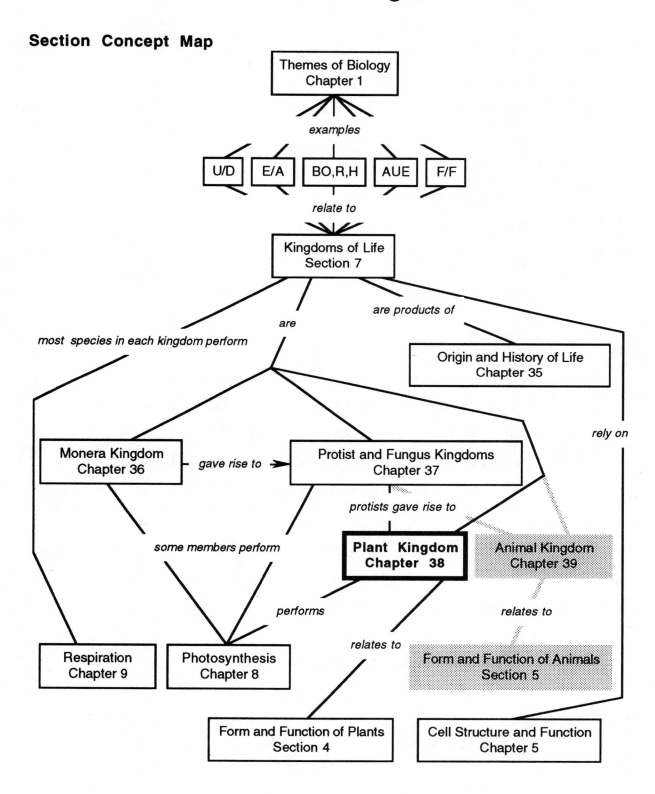

Go Figure!

Figure 38-2: What do the different amounts of colored space in each circle represent?

What process allows the 2n sporophyte to become haploid (1n)?

What process allows the haploid generation to become diploid again?

Figure 38-4: Where do plants alive today appear on the diagram?

Why are two lines leading to the bryophytes?

Figure 38-5: Which diagram shows bryophytes and tracheophytes with green algae as a common ancestor?

Why isn't the hypothesis of bryophytes giving rise to tracheophytes considered as a third option?

Figure 38-6: To which division does each organism belong?

Figure 38-7: *Chlamydomonas navalis* is pink. Why isn't it considered a red alga?

Figure 38-8: With the exception of the objects that look like angry cobras in the moss photograph, all the plants are haploid. Which generation is shown?

What moss generation do the "angry cobras" represent?

Are they haploid or diploid?

Figure 38-9: What structure surrounds the zygote?

What process transforms the zygote into a multicellular sporophyte?

How does the single celled spore develop into a multicellular gametophyte?

Figure 38-10: If the caption hadn't told you this is a sporophyte, how would you have known?

Figure 38-11: In what ways does *Psilotum* resemble the fossil *Rhynia* in Figure 38-10?

Figure 38-15: Observe the part labelled "Developing sporophyte." The vertical plant part is the first sporophyte leaf; the sporophyte root is projecting downwards. What is the horizontal part?

Figure 38-16: What kind of chemical bonds are broken when the sporangium releases its spores?

Figure 38-17: Why are cycads considered gymnosperms when many resemble palms?

Figure 38-20: Is the pine shown monoecious or dioecious?

What process produces microspores?

What process produces megaspores?

Which gamete is contained inside pollen grains?

Figure 38-21: What dicot characteristics are visible in pictures Figures 38-21a and 38-21b?

What monocot characteristics are visible in pictures Figures 38-21c and 38-21d? ("None" is an acceptable answer!)

Matching Questions

Write the letter of the phrase that best matches the numbered term on the left. Use each only once.

_____ 1. sporophyte generation

_____ 2. gametophyte generation

_____ 3. thallus

_____ 4. bryophyta

_____ 5. chlorophyta

_____ 6. fucoxanthin

_____ 7. antheridium

_____ 8. homosporous

_____ 9. megaspore

_____10. strobilus

_____11. frond

_____12. prothallus

_____13. dioecious

_____14. coniferophyta

_____15. sphenophyta

_____16. pterophyta

_____17. epiphyte

_____18. ovule

_____19. microgametophyte

_____20. fruit

a. a flat plant body with little cellular specialization

b. pigment characteristic of phaeophytes

c. develops into a female gametophyte

d. bryophyte sperm producing structure

e. fern leaf

f. horsetails belong to this division

g. made of a megasporangium, other sporophyte tissue, and a female gametophyte

h. develops from a microspore

i. develops from an angiosperm ovary

j. most obvious generation of higher vascular plants

k. fern gametophyte

l. most common gymnosperm division

m. plant group composed of mosses and liverworts

n. a plant that grows on another plant, but is not parasitic

o. describes a plant species that produces only a single type of spore

p. the division of ferns

q. the division of green algae

r. a cluster of sporophylls in certain vascular plants

s. most obvious generation of lower vascular plants

t. male and female parts are on different individuals

Multiple Choice Questions

1. Which is an adaptation of plants to a terrestrial environment?
    a. photosynthesis
    b. fermentation
    c. cuticle
    d. chlorophyll
    e. gametes

2. Which is an adaptation of plants to a terrestrial environment?
    a. xylem and phloem
    b. cellular respiration
    c. chloroplasts
    d. motile sperm
    e. mitochondria

3. Which describes the chlorophyta?
    a. This division includes all algae.
    b. It is thought that an ancient chlorophyte is ancestral to true plants.
    c. Fucoxanthin is the primary photosynthetic pigment.
    d. Most are marine plants.
    e. Members have a peptidoglycan cell wall.

4. Which describes the chlorophyta?
    a. The sporophyte is the dominant generation in most species.
    b. Male and female gametes are indistinguishable in all species.
    c. The pigments chlorophyll *a* and *b* are present in all species.
    d. Glycogen is the storage carbohydrate.
    e. Members are unicellular or colonial, but never multicellular.

5. Which describes the phaeophyta?
    a. Growth rates of all species are very slow.
    b. The plants tend to grow in shallow water.
    c. There is no cellular differentiation in any species.
    d. Species have either a sporophyte generation or a gametophyte generation, never both.
    e. This is the division of brown algae.

6. Which describes the phaeophyta?
    a. The group is important ecologically, but has no commercial importance.
    b. Large species may have holdfasts, blades and floats.
    c. The gametophyte is the obvious generation of large species.
    d. The most important accessory pigment is rhodopsin.
    e. Some plants are the source of agar.

7. Which describes the rhodophyta?
    a. It is thought that an ancient rhodophyte is ancestral to true plants.
    b. The members are predominantly found in freshwater environments.
    c. They are important members of the large kelp forests.
    d. Because of their pigments, they are able to survive in deeper waters.
    e. They are commonly called brown algae.

8. Which describes the rhodophyta?
    a. This is the division of green algae.
    b. The members have chlorophyll *a* and *c*.
    c. No sexual stage in the life cycle has been discovered as yet.
    d. No sporophyte stage in the life cycle has been discovered as yet.
    e. Most members live in terrestrial habitats.

9. Which describes the bryophyta?
   a. Bryophytes have no cellular specialization.
   b. Bryophytes one meter or more in height are common.
   c. Bryophytes exhibit alternation of generations.
   d. Bryophyte seeds are capable of remaining dormant for long periods of time.
   e. Bryophytes are not capable of sexual reproduction.

10. Which describes the bryophyta?
    a. The diploid spore producing generation is most obvious.
    b. Water is necessary for bryophyte sperm to swim in on their journey to the egg.
    c. The gametophyte is dependent on the sporophyte for nutrition.
    d. Rhizoids are specialized for photosynthesis.
    e. Taller bryophytes have xylem and phloem.

11. During a morning stroll on the beach, you discover a large brownish plant with leafy structures washed up onto shore. You take it back to your laboratory and discover it has fucoxanthin pigment in its cells. To which division does this plant belong?
    a. bryophyta
    b. chlorophyta
    c. phaeophyta
    d. rhodophyta
    e. xanthophyta

12. While exploring in a forest, you find a small greenish plant growing on a rock near a stream. Microscopic examination reveals haploid cells and multicellular gametangia. Chemical analysis indicates starch and chlorophyll *a* and *b* are present. To which division does this plant belong?
    a. bryophyta
    b. chlorophyta
    c. phaeophyta
    d. rhodophyta
    e. xanthophyta

13. Which describes the lycophyta?
    a. It is a division of lower vascular plants.
    b. All members of the division are homosporous.
    c. Megagametophytes have sporophylls for spore production.
    d. Modern members of the division do not require water for fertilization.
    e. Each strobilus contains a single seed.

14. Which describes the lycophyta?
    a. Modern members of the division may be tall, woody plants.
    b. All members of the division are heterosporous.
    c. Heterosporous species produce megagametophytes and microgametophytes.
    d. Microgametophytes have sporophylls for spore production.
    e. Each strobilus contains many seeds.

15. Which describes the pterophyta?
    a. Most members of the division are heterosporous.
    b. Sporangia are grouped into sori.
    c. They do not require water for fertilization.
    d. Sporangia are found on the prothallus.
    e. The gametophyte generation is most obvious.

16.   Which describes the pterophyta?
    a.   They were the first vascular plants to produce seeds.
    b.   Gametophytes have vascular tissue.
    c.   Insects play an important role in transfer of spores between plants.
    d.   The most obvious stage of the life cycle is diploid.
    e.   This is the division of whisk ferns.

17.   Which is a gymnosperm advancement over lower vascular plants?
    a.   Gymnosperms protect sperm cells in a dry pollen grain.
    b.   Gymnosperms protect the embryo in a fruit.
    c.   Gymnosperms produce xylem and phloem.
    d.   Gymnosperms produce roots, stems and leaves.
    e.   Gymnosperms produce spores.

18.   Which is a gymnosperm advancement over lower vascular plants?
    a.   Gymnosperms produce gametes.
    b.   Gymnosperms produce motile sperm.
    c.   Gymnosperms are able to perform cellular respiration.
    d.   Gymnosperms are able to photosynthesize.
    e.   Gymnosperms protect the embryo in a seed.

19.   Which is a gymnosperm division?
    a.   psilophyta
    b.   cycadophyta
    c.   anthophyta
    d.   chrysophyta
    e.   pyrrophyta

20.   Which is a gymnosperm division?
    a.   ginkgophyta
    b.   sphenophyta
    c.   chlorophyta
    d.   cyanophyta
    e.   tracheophyta

21.   Which describes pines?
    a.   They belong to the smallest gymnosperm division.
    b.   Most species are dioecious.
    c.   All species are heterosporous.
    d.   Pollen is produced in the ovule.
    e.   It takes a single growing season to go from production of a female cone to seed formation.

22.   Which describes pines?
    a.   Each ovule contains antheridia for sperm production.
    b.   Sperm have two flagella for motility.
    c.   Megaspore mother cells divide to produce male gametophytes.
    d.   Pollen grains have structures that aid in wind dispersal.
    e.   Cones are produced by conifer gametophytes.

23.   Which is an angiosperm advancement over gymnosperms?
    a.   Angiosperms do not require water for fertilization.
    b.   Angiosperms produce flowers to attract specific pollinators.
    c.   Angiosperms produce pollen.
    d.   Angiosperms have a dominant gametophyte generation.
    e.   Angiosperms are heterosporous.

24.   Which is an angiosperm advancement over gymnosperms?
   a.   Angiosperms require water for fertilization.
   b.   Angiosperms produce seeds.
   c.   Angiosperms produce ovules.
   d.   Angiosperms have a dominant sporophyte generation.
   e.   Angiosperms produce fruit for seed dispersal.

Concept Map Construction

Construct a concept map for each group of terms. Be sure to include appropriate connector phrases.
You may add other terms as necessary and use terms in the singular or plural form.

   1.   alternation of generations, fertilization, chlorophyta, sperm, spore
   2.   psilophyta, pterophyta, xylem, sorus, meiosis
   3.   coniferophyta, monocot, ovule, cone, sporophyte

Chapter 39

The Animal Kingdom

Section Concept Map

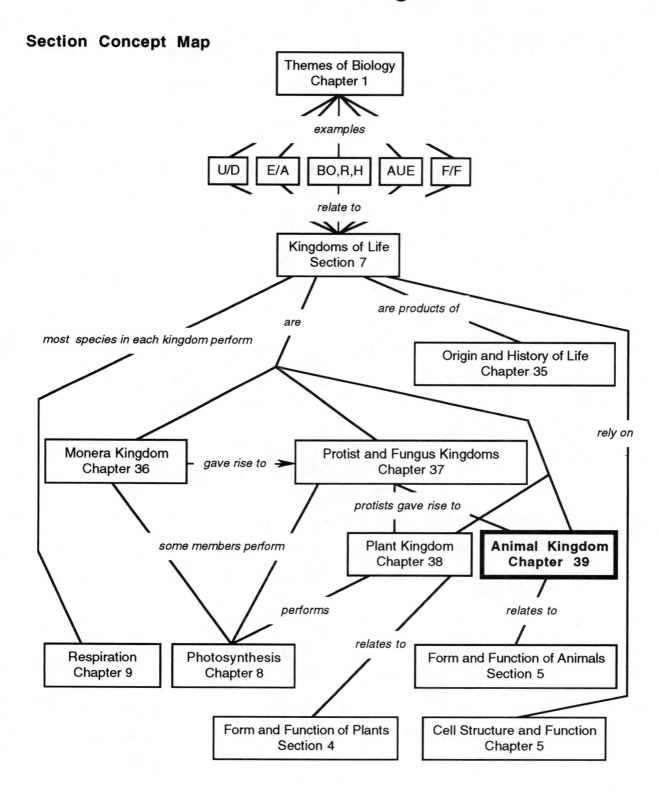

Go Figure!

Figure 39-1: Why are such obviously different organisms as *Trichoplax adhaerans* and a blue whale considered members of the same kingdom?

At what taxonomic level are the two separated?

Figure 39-3: Since the elephant can be divided into three basic planes (median, transverse, and frontal), why do biologists consider it to have bilateral symmetry?

Figure 39-5: What is the significance of arthropods and segmented worms coming off the same branch after splitting with mollusks?

According to this diagram, which arose first: acoelomates or pseudocoelomates? ...crustaceans or arachnids? ...echinoderms or fish?

What does the segment between the branch leading to modern amphibians and the branch leading to modern reptiles represent?

Figure 39-6: What is the function of porocytes?

What is the function of choanocytes?

How is increased surface area for choanocytes an advantage?

Figure 39-7: Which organisms are polyps?

Are any in the medusa form?

What is the fundamental difference between polyps and medusas?

What kind of symmetry does each have?

Figure 39-8: Why are cnidarians classified as only having a tissue level of organization?

What are those tissues?

Where are cnidocytes concentrated in each body form?

What is the role of cnidocytes?

Figure 39-11: What kind of symmetry does the flatworm have?

What two structures shown give evidence of cephalization?

Figure 39-12: Based on the drawings in Figure 39-11, what does the central opening represent in this drawing?

What do the smaller openings on either side represent?

Figure 39-13: How many different organ systems are represented in this "simple" worm?

How does a pseudocoelom differ from a true coelom?

Figure 39-15: How many different organ systems are represented in this drawing of a snail?

Which characteristics are distinctly molluscan?

Figure 39-17: Which two advancements made by annelids are shown in this picture?

How many different organ systems are represented?

What evidence is there of cephalization?

Figure 39-18: What arthropod characteristics can you actually see in each of these photographs?

How is it possible for biologists to look at a fossil of an extinct organism (such as the trilobite) and assign it with reasonable certainty to a taxonomic group (such as the phylum Arthropoda)?

Figure 39-19: What features does the grasshopper have that humans also have?

What features do humans have that grasshoppers don't?

What features do grasshoppers have that humans don't?

Figure 39-20: What visible features justify placing these diverse organisms in the same taxonomic class?

Figure 39-23: What echinoderm features are visible in both organisms?

Figure 39-24: What echinoderm features are visible in this sea star diagram?

How many organ systems are represented?

Is there any evidence of cephalization or segmentation?

Figure 39-25: What kind of symmetry is exhibited by this generalized chordate?

What evidence is there of cephalization?

What evidence is there of segmentation?

Is there a coelom?

Figure 39-26: If the adult tunicate lacks most chordate characteristics, why are tunicates considered chordates?

Figure 39-27: Compare this diagram with the generalized chordate in Figure 39-25. Why do you suppose they are so similar?

Figure 39-28: What part of the lamprey is shown in Figure 39-28a and why aren't lampreys considered radially symmetrical?

Figure 39-29: What advantages are there to hinged jaws?

Figure 39-30: What vertebrate characteristics are visible?

What visible characteristics indicate this is a cartilagenous fish (class Chondricthyes)?

What other cartilagenous fish features can not be seen?

Figure 39-31: What vertebrate characteristics are visible?

What visible characteristics indicate these are bony fish (class Osteichthyes)?

What other bony fish features can not be seen?

Figure 39-32: What vertebrate characteristics are visible?

What visible characteristics indicate these are amphibians (class Amphibia)?

What other amphibian features can not be seen?

Figure 39-33: What vertebrate characteristics are visible?

What visible characteristics indicate these are reptiles (class Reptilia)?

What other reptilian features can not be seen?

Figure 39-34: What vertebrate characteristics are visible in the photograph of the hummingbird?

What visible characteristics indicate it is a bird (class Aves)?

What other avian features can not be seen?

What flight adaptations do birds have?

Figure 39-35: What vertebrate characteristics are visible?

What visible characteristics indicate these are mammals (class Mammalia)?

What other mammalian features can not be seen?

How are the three mammalian groups distinguished?

Figure 39-36: According to the diagram, how many years has it been since humans and tree shrews shared a common ancestor?

How many years has it been since humans and any other modern primate shared a common ancestor?

Some people have the mistaken notion that our closest primate relatives, the great apes, will eventually evolve into humans. How could you use this diagram to correct their misconception?

Matching Questions

Write the letter of the phrase that best matches the numbered term on the left. Use each only once.

_____ 1. radial symmetry

_____ 2. bilateral symmetry

_____ 3. cephalization

_____ 4. body segment

_____ 5. spongocoel

_____ 6. polyp

_____ 7. acoelomate

_____ 8. pseudocoelomate

_____ 9. deuterostome

_____10. radula

_____11. parapodia

_____12. hemocoel

_____13. tube feet

_____14. ostracoderm

_____15. swim bladder

_____16. amphibian

_____17. amniotic egg

_____18. monotreme

_____19. marsupial

_____20. primate

a. unjointed appendages of some annelids

b. a cnidarian body form

c. having a body cavity not lined with epithelium

d. a body plan that can be divided through a central point into many mirror images

e. egg-laying mammal

f. part of the echinoderm water vascular system

g. organ used for buoyancy in bony fish

h. one of repeating subunits in body construction

i. pouched mammal

j. a body plan that can be divided only into right and left halves

k. first vertebrate class to be terrestrial

l. an evolutionary development credited to reptiles

m. animal in which the blastopore becomes the anus

n. mammalian order that includes apes and monkeys

o. cavity characteristic of sponges

p. an extinct jawless fish

q. a cavity filled with blood

r. lacking an internal body cavity

s. a structure containing chitinous teeth

t. evolutionary trend towards developing sense organs and nervous tissue at the front of an animal

Multiple Choice Questions

1. Which IS NOT typical of animals with radial symmetry?
   a. They are capable of moving quickly through their environment.
   b. Sensory receptors are evenly distributed over body surfaces.
   c. Their body can be divided in more than one way to produce mirror images.
   d. Their body is cylindrical.
   e. They experience the environment equally from all directions.

2. Which IS NOT typical of animals with bilateral symmetry?
   a. They demonstrate cephalization.
   b. Their body can be divided in only one plane to produce mirror images.
   c. Their body has a definite front end to it.
   d. They are sessile.
   e. Sensory receptors are concentrated in one part of the body.

3. Which describes porifera?
   a. They have bilateral symmetry.
   b. They are predators.
   c. Choanocytes are specialized feeding cells.
   d. They have a tissue level of organization.
   e. They have a chitinous exoskeleton.

4. Which describes porifera?
   a. They have radial symmetry.
   b. They are filter feeders.
   c. Porocytes are cells specialized for motility.
   d. They have an organ level of organization.
   e. They have a calcium carbonate shell.

5. Which describes cnidarians?
   a. They have bilateral symmetry.
   b. They are herbivorous.
   c. All cnidarians go through a polyp stage at some point in the life cycle.
   d. The gastrovascular cavity is lined with epidermis.
   e. They have a tissue level of organization.

6. Which describes cnidarians?
   a. They have some type of radial symmetry.
   b. They are autotrophic.
   c. All cnidarians go through a medusa stage at some point in the life cycle.
   d. Cnidocytes are found only in the gastrodermis.
   e. They have an organ level of organization.

7. Which describes platyhelminths?
   a. They are pseudocoelomates.
   b. They have no symmetry.
   c. They have a brain and nerve cords.
   d. They have a tissue level of organization.
   e. All are phagocytic.

8. Which describes platyhelminths?
   a. They have a true coelom.
   b. They have radial symmetry.
   c. They have no nervous tissue.
   d. They have an organ level of organization.
   e. Some are free-living and others are parasitic.

9. Which describes nematodes?
   a. They are acoelomates.
   b. They have radial symmetry.
   c. They have a cartilagenous skeleton.
   d. They are unsegmented.
   e. They have jointed appendages.

10. Which describes nematodes?
   a. They are pseudocoelomates.
   b. They have no symmetry.
   c. They have spicules for skeletal support.
   d. They are segmented.
   e. They have unjointed appendages.

11. Which DOES NOT describe mollusks?
   a. Their internal organs are concentrated in a dorsal visceral mass.
   b. They belong to the second largest animal phylum.
   c. Their shell is made of protein and calcium carbonate.
   d. They are protostomes.
   e. They are segmented.

12. Which DOES NOT describe mollusks?
   a. Many have a radula with chitinous teeth.
   b. The mantle secretes a shell, if present.
   c. Most have a muscular foot for locomotion.
   d. They are deuterostomes.
   e. They are unsegmented.

13. Which describes annelids?
   a. They are segmented.
   b. They are acoelomates.
   c. They have jointed appendages.
   d. They have a chitinous exoskeleton.
   e. They are deuterostomes.

14. Which describes annelids?
   a. They are unsegmented.
   b. They are pseudocoelomates.
   c. They have no appendages.
   d. Their body cavity acts as a hydrostatic skeleton.
   e. They evolved from a poriferan ancestor.

15. Which DOES NOT describe arthropods?
   a. The hemocoel is filled with blood.
   b. Lungs are found in terrestrial arthropods.
   c. They have a system level of organization.
   d. They have bilateral symmetry.
   e. Molting of the exoskeleton allows growth.

16. Which DOES NOT describe arthropods?
   a. They exhibit the greatest amount of diversity of any animal phylum.
   b. They are protostomes.
   c. Their largest cavity is the coelom.
   d. They have an exoskeleton made of chitin and protein.
   e. They have jointed appendages.

17. Which DOES NOT describe echinoderms?
    a. They have a chitinous exoskeleton.
    b. The larvae exhibit bilateral symmetry.
    c. Their adult body plan is pentamerous.
    d. Some are capable of everting their stomach.
    e. Tube feet are operated by the water vascular system.

18. Which DOES NOT describe echinoderms?
    a. The adults exhibit radial symmetry.
    b. The water vascular system is involved in respiration.
    c. They are deuterostomes.
    d. Their calcium carbonate skeleton has projecting spines.
    e. Because of the similarity in symmetry, echinoderms are thought to have close ancestry to cnidarians.

19. Which is a characteristic unique to chordates?
    a. bilateral symmetry
    b. deuterostomic development
    c. notochord
    d. motility
    e. cephalization

20. Which is a characteristic unique to chordates?
    a. a tail
    b. segmentation
    c. a digestive tube with mouth and anus
    d. pharyngeal gill slits
    e. nervous tissue organized into a brain and nerve cords

21. Which group is characterized by three body segments, six jointed legs, wings, and an exoskeleton?
    a. gastropods
    b. oligochaetes
    c. insects
    d. crustaceans
    e. hirudineans

22. Which group is characterized by a true coelom, segmentation, and unjointed appendages?
    a. polychaetes
    b. cephalopods
    c. nematodes
    d. chelicerates
    e. myriapods

23. Which group is characterized by a mantle cavity, paired shells, and a dorsal visceral mass?
    a. oligochaetes
    b. gastropods
    c. insects
    d. bivalves
    e. platyhelminths

24. Which group is characterized by no coelom, radial symmetry, cnidocytes, and a gastrovascular cavity?
    a. cnidarians
    b. gastropods
    c. cephalopods
    d. echinoderms
    e. crustaceans

25.   Which vertebrate class is characterized by gills, swim bladder and a bony skeleton?
      a.   chondrichthyes
      b.   osteichthyes
      c.   amphibia
      d.   reptilia
      e.   aves

26.   Which vertebrate class is terrestrial but must return to an aquatic environment to reproduce?
      a.   chondrichthyes
      b.   mammalia
      c.   amphibia
      d.   reptilia
      e.   aves

27.   Which vertebrate class is endothermic, has fur and nurses their young?
      a.   mammalia
      b.   osteichthyes
      c.   amphibia
      d.   reptilia
      e.   aves

28.   Which vertebrate class is characterized by a dry, scaly skin, and the production of an amniotic egg?
      a.   chondrichthyes
      b.   osteichthyes
      c.   amphibia
      d.   reptilia
      e.   mammalia

29.   Which vertebrate class is characterized by a cartilagenous skeleton and denticles?
      a.   chondrichthyes
      b.   osteichthyes
      c.   amphibia
      d.   reptilia
      e.   aves

30.   Which vertebrate class produces an amniotic egg and is endothermic?
      a.   chondrichthyes
      b.   osteichthyes
      c.   amphibia
      d.   mammalia
      e.   aves

Concept Map Construction

Construct a concept map for each group of terms. Be sure to include appropriate connector phrases.
You may add other terms as necessary and use terms in the singular or plural form.

1.   radial symmetry, segmentation, invertebrate, coelom, spongocoel
2.   protostome, appendage, insect, bivalve, exoskeleton
3.   amniotic egg, endothermic, chordate, echinoderm, cnidaria

Chapter 40

The Biosphere

Section Concept Map

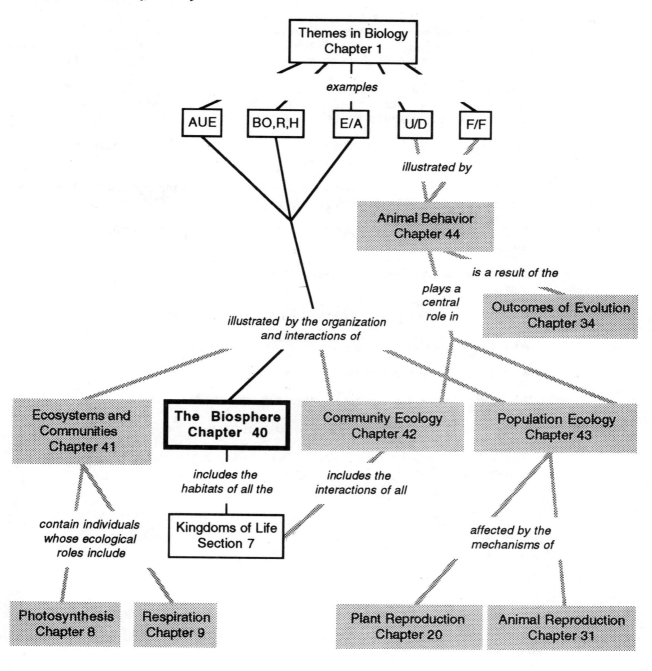

Themes in Biology
Chapter 1

examples

AUE   BO,R,H   E/A   U/D   F/F

illustrated by

Animal Behavior
Chapter 44

is a result of the

plays a central role in

Outcomes of Evolution
Chapter 34

illustrated by the organization and interactions of

Ecosystems and Communities
Chapter 41

The Biosphere
Chapter 40

Community Ecology
Chapter 42

Population Ecology
Chapter 43

contain individuals whose ecological roles include

includes the habitats of all the

includes the interactions of all

Kingdoms of Life
Section 7

affected by the mechanisms of

Photosynthesis
Chapter 8

Respiration
Chapter 9

Plant Reproduction
Chapter 20

Animal Reproduction
Chapter 31

Go Figure!

Figure 40-2(d):   Which line represents the equator?...the tropic of Cancer?...the tropic of Capricorn?

The three parallel lines represent the north to south range of tropical latitudes. What physical characteristic determines this range? Are tropical rain forests restricted to this region? Which biomes are not generally found in this region?

Figure 40-7:   During low tide, water trapped in depressions is cut off from the ocean. What might happen to the physical conditions in these small pools while the tide is out? How would this affect temperature?...salt content of the water?...oxygen content of the water? How might changes in these conditions affect organisms left in the pools by the receding tide?

Most organisms living in these intertidal habitats reside in or upon the substrate. How does the stability of bottom substrate differ between these habitats? How might these differences affect the number and type of organisms living there?

Figure 40-8:   Describe potential environmental temperature differences between organisms living in the splash zone and those in the low intertidal zone.

Two filter feeding species live in the intertidal habitat, one in the high intertidal the other in the middle intertidal zone. How would this difference affect food availability?

Figure 40-10:   What does the red line represent? How does this line change from spring to summer to autumn?

What does the green shading represent? How does this condition change between seasons?

What does the darkest green shaded area under "Summer" represent?

What do the green arrows represent under "Spring and autumn?"

Why is warmer water more buoyant?

What forces drive spring overturn?...autumn overturn?

Figure 40-12:   Is it possible to have tundra at the equator according to this figure?

What will happen to the elevation of tree line as you move north from the equator?...the elevation at which coniferous forest occurs?

Matching Questions

Write the letter of the phrase that best matches the numbered term on the left. Use each only once.

_____ 1. CFC	a. defined as...includes living and non-living components of the environment
_____ 2. hydrosphere	
_____ 3. oikos	b. formed by air moving from 30° latitude toward 60° latitude
_____ 4. ecosystem	
_____ 5. climate	c. dominated by evergreen trees
_____ 6. trade winds	d. covered by water 50% of the time
_____ 7. westerlies	e. permanently frozen soil
_____ 8. salinity	f. home
_____ 9. epipelagic zone	g. refers to sea floor
_____10. benthic zone	h. formed by air moving from 30° latitude toward the equator
_____11. plankton	
_____12. upwelling	i. found at high elevation above tree line
_____13. fringe reef	j. chlorofluorocarbon
_____14. middle intertidal zone	k. upper layer of a tropical rain forest
_____15. limnetic zone	l. surfacing of nutrient-ladened water
_____16. shrubland	m. photic zone
_____17. coniferous forest	n. a lake's open, lighted waters
_____18. overstory	o. water subdivision of the biosphere
_____19. alpine tundra	p. biome of short, woody plants
_____20. permafrost	q. extends from shore out into the ocean
	r. organisms that drift with ocean currents
	s. prevailing weather in an area
	t. water's salt concentration

Multiple Choice Questions

1. The greatest quantity of life (number of individuals) is found in
   a. the hydrosphere.
   b. the lithosphere.
   c. the atmosphere.
   d. the tropical rain forest of the lithosphere.
   e. terrestrial habitats.

2. Where do most organisms of the lithospere reside?
   a. buried deep in the soil
   b. near the illuminated surface
   c. in chaparral
   d. at the edge of the hydrosphere
   e. in the littoral zone

3. What type of organism does your text say is occasionally found in the atmosphere at elevations up to 62 miles?
   a. spider
   b. fly
   c. fungal spore
   d. butterfly
   e. sparrow

4. What type of organism does your text say is occasionally found in the atmosphere at elevations up to 62 miles?
   a. wasp
   b. beetle
   c. cockroach
   d. bacterium
   e. virus

5. Based upon the definition of ecology given by Charles Krebs, ecology includes the study of
   a. natural selection.
   b. adaptation.
   c. how many organisms occur in a given place.
   d. the origin of variability within a species.
   e. an organism's ancestral history.

6. Based upon the definition of ecology given by Charles Krebs, ecology includes the study of
   a. development.
   b. why organisms occur where they do.
   c. patterns of inheritance.
   d. origin of homologous structures.
   e. origin of island chains.

7. The earth's climates are primarily affected by the circulation patterns of
   a. the lithosphere.
   b. the pelagic zone.
   c. the atmosphere.
   d. the bathypelagic zone.
   e. phytoplankton.

8. The earth's climates are primarily affected by the circulation patterns of
    a. acid rain.
    b. estuaries.
    c. the epipelagic zone.
    d. the oceans.
    e. the profundal zone.

9. During its summer, the Northern Hemisphere
    a. loses less heat than the Southern Hemisphere.
    b. is tipped towards the sun.
    c. is tipped away from the sun.
    d. gains heat from the Southern Hemisphere.
    e. is closer to the moon.

10. During its winter, the Northern Hemisphere
    a. loses less heat than the Southern Hemisphere.
    b. is tipped towards the sun.
    c. is tipped away from the sun.
    d. gains heat from the Southern Hemisphere.
    e. is closer to the moon.

Use the following information for questions 11 and 12.

Rising air creates areas of low atmospheric pressure. Areas of sinking air have high atmospheric pressure.

11. Which latitudes have prevailing high pressure?
    a. 0° and 30° north
    b. 30° north and 30° south
    c. 30° south and 60° north
    d. 60° north and 60° south
    e. 60° north and 90° north

12. Which latitudes have prevailing low pressure?
    a. 0° and 30° north
    b. 30° north and 30° south
    c. 30° south and 60° north
    d. 60° north and 60° south
    e. 60° north and 90° north

13. Neritic waters are generally very rich in life. Which is a factor important in allowing such species diversity?
    a. Neritic areas are generally very warm.
    b. All neritic waters surround coral reefs, and most reef species spend at least part of their life cycle there.
    c. They generally contain abundant nutrients from the neighboring land.
    d. They are generally very calm, non-turbulent areas.
    e. Hydrothermal vents provide neritic waters with an abundance of nutrients.

14. Neritic waters are generally very rich in life. Which is a factor important in allowing such species diversity?
    a. Neritic areas are generally very cold.
    b. All neritic waters surround estuaries, and most estuarine species spend at least part of their life cycle there.
    c. They generally are nutrient poor areas.
    d. Turbulence stirs the water and distributes nutrients to photosynthetic organisms.
    e. Coral reefs provide neritic waters with an abundance of nutrients.

15. Which is an accurate comparison of a lake's limnetic and profundal zones?
    a. The limnetic zone is the area around the lake's edge, and the profundal zone is all other lighted areas.
    b. The limnetic zone is the area of the lake's open, lighted water, and the profundal zone is all other lighted areas.
    c. The limnetic zone is the area of the lake's open, lighted water, and the profundal zone is the area below the depth light penetrates.
    d. The limnetic zone is the area around the lake's edge, and the profundal zone is the area below the depth light penetrates.
    e. The profundal zone is the area around the lake's edge, and the limnetic zone is the area below which light penetrates.

16. Which is an accurate comparison of a lake's limnetic and profundal zones?
    a. Phytoplankton are the major photosynthetic producers of the limnetic zone, and rooted plants are the major photosynthetic producers of the profundal zone.
    b. Phytoplankton are the major photosynthetic producers of the limnetic zone, and there are no photosynthetic producers in the profundal zone.
    c. Rooted plants are the major photosynthetic producers of the limnetic zone, and phytoplankton are the major photosynthetic producers of the profundal zone.
    d. Floating multicellular algae are the major photosynthetic producers of the limnetic zone, and there are no photosynthetic producers in the profundal zone.
    e. Floating multicellular algae are the major photosynthetic producers of the profundal zone, and there are no photosynthetic producers in the limnetic zone.

17. Which biome supports the greatest species diversity?
    a. tropical rain forest
    b. deciduous forest
    c. coniferous forest
    d. grassland
    e. tundra

18. Which biome supports the least species diversity?
    a. tropical rain forest
    b. deciduous forest
    c. coniferous forest
    d. grassland
    e. tundra

19. In what way are tropical rain forests and coniferous forests similar?
    a. They have a comparable rate of decomposition.
    b. Both typically have three stories of plants.
    c. Both have great species diversity.
    d. Both have relatively nutrient poor soil.
    e. Over half of all species on earth are evenly divided between these two biomes.

20. In what way are tropical rain forests and coniferous forests similar?
    a. They have a comparable rate of leaf litter accumulation.
    b. Both typically have two stories of plants.
    c. Both have little species diversity.
    d. Both have relatively nutrient rich soil.
    e. Over one third of all species on earth are evenly divided between these two biomes.

21. Which biome is found in regions with a Mediterranean climate and is characterized by woody, perennial plants, shorter than trees with small, leathery leaves?
    a. tundra
    b. coniferous forest
    c. grassland
    d. shrubland
    e. desert

22. Which biome comprises 30 percent of the earth's terrestrial surface and occurs in areas with descending air?
    a. tundra
    b. coniferous forest
    c. grassland
    d. shrubland
    e. desert

Concept Map Construction

Construct a concept map for each group of terms. Be sure to include appropriate connector phrases. You may add other terms as necessary and use terms in the plural or singular form.

1. osmoregulation, intertidal zone, low intertidal zone, estuary, filter feeder
2. photosynthesis, limnetic zone, mesotrophic, species diversity, salinity
3. ecosystem, biome, intertidal zone, stabilizing selection, tropical rain forest

Chapter 41

Ecosystems And Communities

Section Concept Map

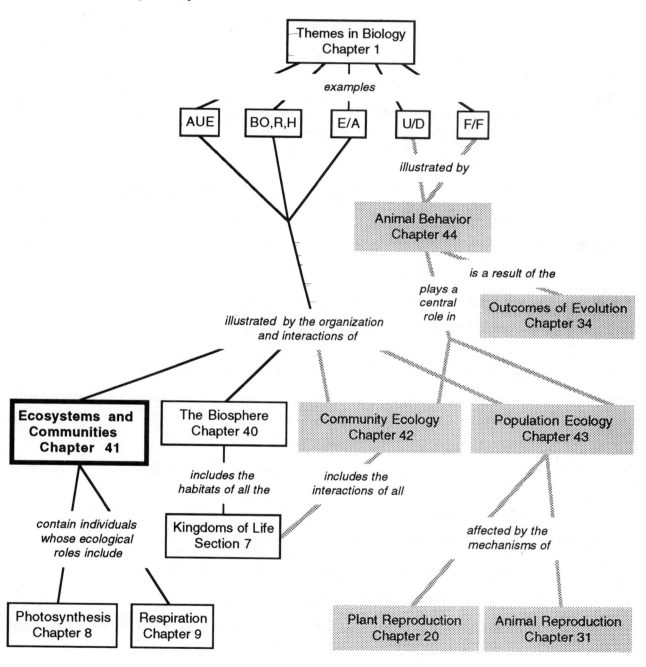

Go Figure!

Figure 41-4:   What would happen to developing frog eggs if environmental temperature went below 0°C?

What would happen to developing frog eggs if environmental temperature was constant at 5°C?

How would the rate of frog egg development compare at 8.75°C and 27°C?

What do tolerance ranges suggest about the environment of trout compared to that of frogs?

Figure 41-6:   Turn your attention to the plot of two factors. If this organism is in a habitat with a temperature of 17°C and a yearly total rainfall of 62 millimeters, will it survive? Will it survive in a habitat with a temperature of 32°C and a yearly total rainfall of 51 millimeters? If a plot of temperature and yearly rainfall produces a point within the yellow area, will the organism live or die?.. within the white area?

Turn your attention to the plot of three factors. If a plot of temperature and yearly rainfall produces a point within the yellow area, will the organism live or die?...outside the yellow area?

Figure 41-8:   Which trophic level contains autotrophic organisms?

Which trophic levels contain heterotrophic organisms?

What is the source of energy used in the first trophic level?

What is the source energy used in the second trophic level?

Figure 41-9:   Five primary consumers are listed for manzanita. In what ways do their niches differ? In what ways are they similar? Since they all feed on manzanita, are they in competition?

Figure 41-10:   When the snake eats frogs, what level of consumer does it represent? When it eats field mice...?

What level of consumer is the field mouse?

A field mouse will also eat grasshoppers. When it does, what level of consumer does it represent?

Figure 41-13:   Which organisms add carbon dioxide to the atmosphere? Which remove it?

How does carbon dioxide form bicarbonate in the lake?

Figure 41-14:   What general type of organism plays a role in Cycle A and Cycle B?

How does the element nitrogen get passed from grass to cow?

Photosynthesis and respiration are antagonistic processes in the carbon cycle. What processes are antagonistic in the nitrogen cycle?

Matching Questions

Write the letter of the phrase that best matches the numbered term on the left. Use each only once.

_____ 1.   ecosystem

_____ 2.   abiotic

_____ 3.   primary producer

_____ 4.   detritus

_____ 5.   Victor Shelford

_____ 6.   ecotypes

_____ 7.   Justus Liebig

_____ 8.   habitat

_____ 9.   fundamental niche

_____10.   realized niche

_____11.   ecological equivalents

_____12.   primary consumer

_____13.   pyramid of energy

_____14.   biogeochemical cycles

_____15.   nitrogen fixation

_____16.   nitrifying bacteria

_____17.   denitrification

_____18.   primary succession

_____19.   pioneer community

_____20.   secondary succession

a.   generally credited with first stating the Law of the Minimum

b.   small disintegrated pieces of organic wastes and dead organisms

c.   term which applies specifically to the physical location of an organism

d.   illustrates the rate at which energy moves through trophic levels

e.   convert ammonia into nitrites

f.   a progression of distinct communities in a previously disturbed habitats

g.   first community in a new habitat

h.   refers to all inanimate parts of an ecosystem

i.   a progression of distinct communities in a new habitat

j.   an organism which feeds upon producers

k.   includes the community and the physical environment

l.   elements passing back and forth between abiotic and biotic ecosystem components

m.   an organism's potential niche

n.   an organism's actual niche

o.   generally credited with first stating the Theory of Tolerance

p.   release of molecular nitrogen from nitrogen compounds

q.   different species with similar niches within any trophic level

r.   conversion of atmospheric nitrogen to ammonia

s.   a population with a genetically fixed tolerance range

t.   autotroph

Multiple Choice Questions

1. The primary ecological levels of organization include
   a. populations.
   b. organisms.
   c. organs.
   d. cells.
   e. molecules.

2. The primary ecological levels of organization include
   a. organ systems.
   b. tissues.
   c. subatomic particles.
   d. ecosystems.
   e. organelles.

3. The relationship between lake and cave ecosystems described in your text illustrates
   a. that ecosystems may function as separate units in terms of their energy use.
   b. that ecosystems may function as separate units in terms of their nutrient use.
   c. that ecosystems may depend on other ecosystems for exchange of materials and energy.
   d. that ecosystems may depend on other ecosystems for exchange of materials only.
   e. that the entire earth is one large ecosystem.

4. The description of a lake and the surrounding terrestrial ecosystem in your text illustrates
   a. that ecosystems may function as separate units in terms of their energy use.
   b. that ecosystems may function as separate units in terms of their nutrient use.
   c. that ecosystems may depend on other ecosystems for exchange of materials and energy.
   d. that ecosystems may depend on other ecosystems for exchange of materials only.
   e. that the entire earth is one large ecosystem.

5. Which is an example of an abiotic component affecting a biotic one?
   a. photosynthesis releasing molecular oxygen into the atmosphere
   b. detritus being added to soil
   c. a coral reef modifying ocean current direction
   d. soil phosphorus content limiting photosynthetic rate
   e. plant roots reducing the rate of soil erosion

6. Which is an example of a biotic component affecting an abiotic one?
   a. molecular oxygen content of soil affecting the abundance of soil nematodes
   b. a forest modifying the local humidity range
   c. ocean currents distribute larval forms along a coast
   d. soil iron content limiting the rate of photosynthetic rate
   e. soil erosion reducing soil fertility

7. The primary ecological role played by algae is
   a. primary producer.
   b. primary consumer.
   c. secondary consumer.
   d. decomposer.
   e. detritovore.

8. The primary ecological role played by nematodes is
   a. primary producer.
   b. primary consumer.
   c. secondary consumer.
   d. decomposer.
   e. detritovore.

9. Trout eggs will not develop if the temperature rises above 12°C. This illustrates
   a. the Theory of Tolerance.
   b. the Law of the Minimum.
   c. the full range of niche breadth.
   d. the existence of ecological equivalents.
   e. biological magnification.

10. When nitrogen concentrations fall below a certain level, phytoplankton's photosynthetic rate is directly related to nitrogen concentration in the water. This illustrates
    a. the Theory of Tolerance.
    b. the Law of the Minimum.
    c. the full range of niche breadth.
    d. the existence of ecological equivalents.
    e. biological magnification.

11. The dry, hot summers of Southern California result in a coastal chaparral biome made up of woody, drought resistant shrubs. This illustrates
    a. the Theory of Tolerance.
    b. the Law of the Minimum.
    c. the full range of niche breadth.
    d. the existence of ecological equivalents.
    e. biological magnification.

12. Coral reefs will not grow in water which falls below 16-18°C. This illustrates
    a. the Theory of Tolerance.
    b. the Law of the Minimum.
    c. the full range of niche breadth.
    d. the existence of ecological equivalents.
    e. biological magnification.

Use the following information for questions 13 and 14.

Barnacle larvae of the genus *Chthamalus* settle and metamorphose to adults over a wide intertidal range. When barnacles of the species *Balanus* are absent from the environment, *Chthamalus* adults survive within the entire intertidal range. When *Balanus* is present, *Chthamalus* adults are found only in the high intertidal zone.

13. Which statement is consistent with these observations?
    a. *Chthamalus* reduces *Balanus'* hypervolume.
    b. The presence of *Balanus* reduces the realized niche of *Chthamalus*.
    c. *Chthamalus* and *Balanus* have identical potential niches.
    d. *Chthamalus* and *Balanus* are ecological equivalents
    e. *Chthamalus* and *Balanus* are ecotypes.

14. Which statement is consistent with these observations?
    a. *Chthamalus* and *Balanus* show niche overlap.
    b. The presence of *Chthamalus* reduces the realized niche of *Balanus*.
    c. *Chthamalus* and *Balanus* form a food chain.
    d. This situation demonstrates ecological succession.
    e. Communities containing *Chthamalus* and *Balanus* are climax communities.

15. Which moves through a food chain?
    a. nutrients
    b. primary producers
    c. guilds
    d. hypervolumes
    e. succession

16. Which moves through a food chain?
    a. primary consumers
    b. detritovores
    c. energy
    d. sunlight
    e. geochemical cycles

17. Under which conditions would you expect to see an ecological pyramid of numbers with a smaller number of producers than primary consumers?
    a. in an ecosystem with small producers and large primary consumers
    b. in an ecosystem with small producers and small primary consumers
    c. in an ecosystem with large producers and large primary consumers
    d. in an ecosystem with large producers and small primary consumers
    e. in an ecosystem with grasses being eaten by mice

18. Under which conditions would an ecological pyramid of biomass have a smaller biomass of producers than primary consumers?
    a. *produce*–large, bottom dwelling algae with no seasonal change in abundance
       primary consumer–small and seasonal in abundance
    b. *producer*–large, bottom dwelling algae with no seasonal change in abundance
       primary consumer–large and seasonal in abundance
    c. *producers*–small, bottom dwelling algae with no seasonal change in abundance
       primary consumers–small and seasonal in abundance
    d. *producers*–small and have significant changes in seasonal abundance
       primary consumers–large with no seasonal changes in abundance except for migrations
    e. *producers*–small and have large changes in seasonal abundance
       primary consumers–large with seasonal changes in abundance excluding migrations

19. How do the carbon and nitrogen biogeochemical cycles compare?
    a. Both use photosynthesis to move atoms between biotic and abiotic components of an ecosystem.
    b. Both directly use aerobic respiration to move atoms between biotic and abiotic components of an ecosystem.
    c. Both have an atmospheric reservoir of element.
    d. Both are dependent upon bacteria and cyanobacteria to move the element between biotic and abiotic components of the cycle.
    e. One has a gaseous phase to the cycle, the other does not.

20. How do the phosphorus and nitrogen biogeochemical cycles compare?
    a. Both use photosynthesis to move atoms between biotic and abiotic components of an ecosystem.
    b. Both use respiration to move atoms between biotic and abiotic components of an ecosystem.
    c. Both have a large atmospheric reservoir of the element.
    d. Both are dependent upon bacteria and cyanobacteria to move the element between biotic and abiotic components of the cycle.
    e. One has a gaseous phase to the cycle, the other does not.

21. How do climax and pioneer communities compare?
    a. Both communities remain constant in species composition for long periods of time.
    b. Both communities have a large net accumulation of biomass.
    c. The plant species in the climax community are generally longer lived than in the pioneer community.
    d. Pioneer communities tend to have more complex food webs than climax communities.
    e. Producers in both communities use a large amount of their energy intake for reproduction.

22. How do climax and pioneer communities compare?
   a. The climax community remains constant in species composition, the pioneer community does not.
   b. Only the climax community has a large net accumulation of biomass.
   c. The plant species in the pioneer community are generally longer lived than in the climax community
   d. Pioneer communities tend to have more primary consumers than climax communities.
   e. Only the climax community autotrophs use a large proportion of their energy intake for reproduction.

Concept Map Construction

Construct a concept map for each group of terms. Be sure to include appropriate connector phrases. You may add other terms as necessary and use terms in the plural or singular form.

1. ecosystem, niche, guild, ecological equivalent, food chain
2. photosynthesis, hydrologic cycle, phosphorus cycle, detritovore, food web
3. biome, pioneer community, species diversity, biomass, tolerance range

Chapter 42

Community Ecology:
Interactions Between Organisms

Section Concept Map

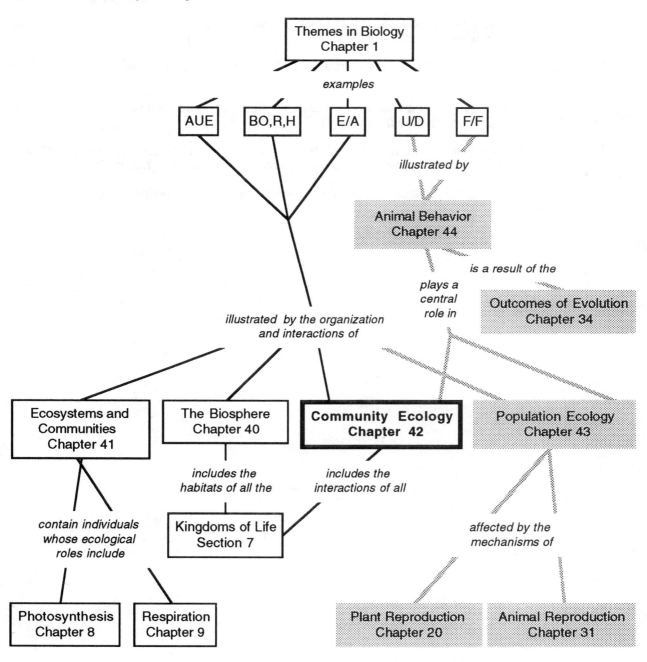

Go Figure!

Figure 42-2:   Compare the diagram on the left with the one on the right. Competition between species A and B is greater on the right than the left. In what other ways do these two graphs differ? Could competition between these two species increase without the other differences?

Figure 42-3:   Describe how the food resource of the Bay-breasted warbler differs from that of the Yellow-rumped warbler.

Describe how food resources of Blackburnian and Bay-breasted warblers differ. What type of resource partitioning is this?

Figure 42-4:   On the average, how many years after the hare population peaked did the lynx population peak?

On the average, how many years after the hare population bottomed out did the lynx population reached their minimum?

What might account for the fact that between 1900 and 1906 hares reached their lowest peak for the entire sample period, but lynx reached their third highest peak?

What was the greatest number of hare pelts taken during this period?...the smallest?

Matching Questions

Write the letter of the phrase that best matches the numbered term on the left. Use each only once.

_____ 1.   serial endosymbiont theory

_____ 2.   herbivore

_____ 3.   interspecific competition

_____ 4.   niche overlap

_____ 5.   interference competition

_____ 6.   resource partitioning

_____ 7.   prey

_____ 8.   carnivore

_____ 9.   camouflage

_____ 10.   escape response

_____ 11.   aposematic coloring

_____ 12.   allelochemical

_____ 13.   mimicry

_____ 14.   Batesian mimicry

_____ 15.   parasite

_____ 16.   parasitoid

_____ 17.   social parasitism

_____ 18.   autotoxicity

_____ 19.   counteradaptation

_____ 20.   commensalism

a.   plant eater

b.   term which specifically denotes interactions between members of different species

c.   dividing an environmental resource between two species

d.   shape or coloration which causes an organism to blend in with its environment

e.   compound which discourages predators

f.   an allelochemical killing the plant that produced it

g.   feeds on a host

h.   results in interspecific competition

i.   describes the origin of eukaryotic organelles

j.   an interaction in which one organism profits and the other is unaffected

k.   specific term which applies to a palatable mimic resembling a distasteful model

l.   organism that eats animals

m.   an adaptation in response to another organisms adaptation

n.   exploiting the behavior of the host

o.   acts as a warning

p.   loss of a lizard's tail is an example

q.   kills its host

r.   general term for any mechanism which increases the chance for survival by one organism resembling another

s.   directly blocking access of another species to environmental resources is an example

t.   an organism fed upon by another organism

Multiple Choice Questions

1. Which is an example of symbiosis?
   a. competition
   b. predation
   c. commensalism
   d. herbivory
   e. competitive exclusion

2. Which IS NOT an example of symbiosis?
   a. parasitism
   b. mimicry
   c. protocooperation
   d. endosymbiosis
   e. mutualism

3. The more similar the niche overlap of two species
   a. the more likely they are to enter into a symbiotic relationship.
   b. the more likely they are to show interspecific competition.
   c. the more likely they are to show intraspecific competition.
   d. the less likely they are to show interspecific competition.
   e. the less likely they are to show intraspecific competition.

4. Which is likely to be the most intense competitive situation?
   a. intraspecific
   b. interspecific
   c. commensalistic
   d. exploitative
   e. allelopathy

5. Your text describes deep-rooted tamarisk trees tapping groundwater supplies in the California desert, thereby reducing water supplies to native mesquite and desert willows. This is an example of
   a. intraspecific competition.
   b. exploitative competition.
   c. interference competition.
   d. territoriality.
   e. character displacement.

6. Your text describes hyenas driving vultures away from the remains of a zebra. This is an example of
   a. intraspecific competition.
   b. exploitative competition.
   c. interference competition.
   d. territoriality.
   e. character displacement.

7. Your text discusses the introduction of domestic goats in 1957 to the island of Abingdon in the Galapagos Archipelago. The goats reduced island food supplies suitable for native tortoises, and by 1962, the tortoises were no longer found on the island. This is an example of
   a. allelopathy.
   b. protocooperation.
   c. commensalism.
   d. competitive exclusion.
   e. resource partitioning.

8. Your text describes five species of North American warblers that feed on insects in slightly different parts of the same tree. This is an example of
    a. allelopathy.
    b. protocooperation.
    c. commensalism.
    d. competitive exclusion.
    e. resource partitioning.

9. Two species of nuthatches with partial overlapping niches are discussed in your text. Where their ranges overlap, bill size and coloration are strikingly different. Where ranges do not overlap, bills are remarkable similar. This is an example of
    a. intraspecific competition.
    b. exploitative competition.
    c. interference competition.
    d. territoriality.
    e. character displacement.

10. When two hypothetical species occupy the same area, they show character displacement, apparently in response to competition. When the two species are examined in parts of their ranges where they do not overlap, what are morphological data about this character likely to reveal?
    a. The two species show even greater differences than where they are cohabitants.
    b. The two species show differences similar to those shown where they are cohabitants.
    c. The two species show fewer differences than where they are cohabitants.
    d. They show exploitative competition.
    e. They show cryptic coloration.

11. Your text classifies "playing dead" as
    a. a group escape adaptation.
    b. an individual escape adaptation.
    c. a defense adaptation.
    d. a camouflage escape adaptation.
    e. Mullerian mimicry.

12. Your text classifies aposematic coloration as
    a. a group escape adaptation.
    b. an individual escape adaptation.
    c. a defense adaptation.
    d. a camouflage escape adaptation.
    e. Mullerian mimicry.

13. When a smelt fish detects an approaching predator, it releases warning chemicals into the water. This is
    a. a group escape adaptation.
    b. an individual escape adaptation.
    c. a defense adaptation.
    d. a camouflage escape adaptation.
    e. Mullerian mimicry.

14. A noxious beetle has a bright orange wingcase with black tips. These are very similar in appearance to the wingcases of another noxious beetle. This is an example of
    a. a group escape adaptation.
    b. an individual escape adaptation.
    c. Batesian mimicry.
    d. a camouflage escape adaptation.
    e. Mullerian mimicry.

15. A mimic gains selective advantage by resembling a distasteful or poisonous species. What is the likely consequence if the mimic is neither distasteful nor poisonous, and its population size has been reduced to one-tenth its normal size?
    a. The selective advantage gained by mimicry will be increased.
    b. The selective advantage gained by mimicry will be decreased.
    c. The mimic will experience an increase in predation.
    d. The distasteful model will experience a decrease in predation.
    e. The selective advantage gained by mimicry will not change.

16. A mimic gains selective advantage by resembling a distasteful or poisonous species. What is the likely consequence if the mimic is neither distasteful nor poisonous, and is ten times as abundant as normal?
    a. The selective advantage gained by mimicry will be increased.
    b. The selective advantage gained by mimicry will be decreased.
    c. The mimic will experience a decrease in predation.
    d. The distasteful model will experience a decrease in predation.
    e. The selective advantage gained by mimicry will not change.

17. The remora/shark relationship is cited by your text as demonstrating
    a. commensalism.
    b. parasitism.
    c. mutualism.
    d. protocooperation.
    e. allelopathy.

18. A lichen is cited by your text as demonstrating
    a. commensalism.
    b. parasitism.
    c. mutualism.
    d. protocooperation.
    e. allelopathy.

19. How does commensalism differ from protocooperation?
    a. In commensalism, one organism benefits and the other is harmed, while in protocooperation both are harmed.
    b. In commensalism, one organism benefits and the other is harmed, while in protocooperation both benefit.
    c. In commensalism, both organisms benefit, while in protocooperation both are harmed.
    d. In commensalism, one organism benefits and the other is unaffected, while in protocooperation both benefit.
    e. In commensalism, both organisms benefit, while in protocooperation one organism benefits.

20. How do mutualism and protocooperation compare?
    a. In mutualism one organism benefits and the other is harmed, while in protocooperation both are harmed.
    b. In mutualism one organism benefits and the other is harmed, while in protocooperation both benefit.
    c. In mutualism both organisms benefit, while in protocooperation both are harmed.
    d. In mutualism one organism benefits and the other is unaffected, while in protocooperation both benefit.
    e. In both mutualism and protocooperation, both organisms benefit.

Concept Map Construction

Construct a concept map for each group of terms. Be sure to include appropriate connector phrases. You may add other terms as necessary and use terms in the plural or singular form.

1.   tolerance range, competition, predation, symbiosis, cryptic coloration
2.   coevolution, counter adaptation, exploitative competition, prey, protocooperation
3.   limiting factor, character displacement, Batesian mimicry, predator, parasitoid

Chapter 43

Population Ecology

Section Concept Map

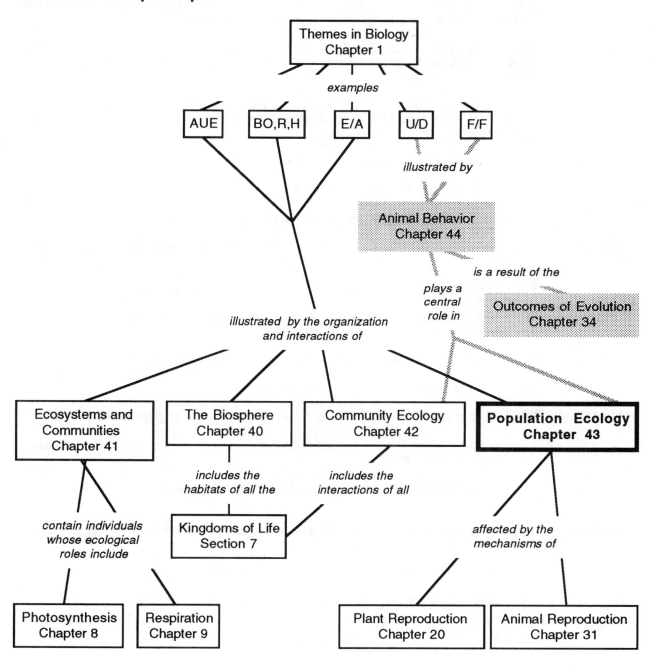

Go Figure!

Figure 43-2:   Please look at Type II survivorship curve. When a population has gone from birth to 20 percent of its life span, what fraction has survived? What is the magnitude of change when a population has gone from 20 to 50 percent of its life span?

Please look at Type III survivorship curve. When a population has gone from birth to 20 percent of its life span, what fraction has survived? What is the magnitude of change when a population has gone from 20 to 50 percent of its life span?

Figure 43-4:   This growth curve is for a population which doubles each time its members synchronously reproduce. How would the curve be affected if population members synchronously had 6 offspring?

This population growth curve is for an organism capable of reproducing every 20 minutes. How would the curve be affected if it was for an organism that had offspring every 6 weeks?

Figure 43-5:   What conditions must exist for dieback #1 to occur?

What conditions might account for dieback #3?

Based upon "b," did this *Daphnia* population go extinct?

Figure 43-6:   What factors may account for the sheep population increasing above average carrying capacity?

What might account for decreases below the carrying capacity average?

What impact would you guess the introduction of sheep had on native Tasmanian herbivores?

Figure 43-7:   Was there any human population growth between 100,000 and 40,000 years ago?

Did the human population grow between 100 and 10 years ago?

Figure 43-8:   What is happening to the length of time it takes to add one billion people to the earth?

What limiting factors kept the human population small early in its history? How have those limiting factors changed?

What has happened to the earth's human carrying capacity between 10,000 BC and the present?

Figure 43-9:   How do these two age-sex structure diagrams differ for prereproductive age groups? How are they similar?

How do they differ for reproductive age groups? How are they similar?

How do they differ for postreproductive age groups? How are they similar?

Matching Questions

Write the letter of the phrase that best matches the numbered term on the left. Use each only once.

_____ 1. population density

_____ 2. population growth rate

_____ 3. fertility rate

_____ 4. sampling

_____ 5. clumped distribution in a population

_____ 6. random distribution in a population

_____ 7. natality

_____ 8. zero population growth

_____ 9. Type I survivorship curve

_____10. biotic potential

_____11. r_o

_____12. $\Delta N/T = r_o N$

_____13. environmental resistance

_____14. K

_____15. dieback

_____16. logistic growth

_____17. r-selected species

_____18. doubling time

_____19. density-dependent limiting factor

_____20. percent annual increase

a. using a small subpopulation to draw conclusion about the whole population

b. birth rate

c. a population's innate capacity to increase

d. number of individuals of a single species per unit area

e. distribution which may result when water is available at a limited number of sites

f. a mortality induced decrease in population size

g. length of time required for population size to increase by a factor of two

h. change in population density per unit of time

i. number of organisms added to a population per 100 individuals in the population

j. sum of all limiting factors

k. results when population mortality is low for most of the species expected life span

l. intrinsic rate of increase

m. produces many offspring at one time

n. a distribution which may result when water is available uniformly in an environment

o. disease

p. is illustrated when a population's growth describes a sigmoidal growth curve

q. rate of exponential growth

r. average number of children born to each reproductive age woman

s. carrying capacity

t. occurs when population gains and losses are equal

Multiple Choice Questions

1. All living organisms within an ecosystem are called
    a. a biome.
    b. a biosphere.
    c. a community.
    d. abiotic conditions.
    e. a population.

2. All of the yellow pine trees in a yellow pine forest comprise
    a. a biome.
    b. a biosphere.
    c. a community.
    d. abiotic conditions.
    e. a population.

3. How do population density and growth rate compare?
    a. Population density is expressed as organisms per unit area and growth rate as clumped or random.
    b. Population density is expressed as change per unit time and growth rate as number per unit area.
    c. Population density is measured as organisms per unit area and growth rate as change per unit time.
    d. Population density is determined directly by natality and growth rate by emigration.
    e. Population density is determined directly by emigration and growth rate by natality.

4. How do population density and distribution compare?
    a. Population density is expressed as organisms per unit area and distribution as clumped or random.
    b. Population density is expressed as change per unit time and distribution as number per unit area.
    c. Population density is measured as organisms per unit area and distribution as change per unit time.
    d. Population density is determined directly by natality and distribution by emigration.
    e. Population density is determined directly by emigration and distribution by natality.

5. A plant population may be spread through the environment in clumps. Which factor probably DOES NOT contribute to this?
    a. Favorable conditions for germination are arranged in patches.
    b. Favorable conditions for photosynthesis are arranged in patches.
    c. Some plant seeds are released in groups.
    d. Some plants reproduce asexually by means of runners.
    e. Some plants release chemicals from their roots which inhibit growth of other members of their species.

6. Some organisms are uniformly spread through the environment. Which factor probably DOES NOT contribute to this?
    a. environmental conditions are highly variable throughout the area
    b. territoriality
    c. release of chemicals from the roots of some plants inhibit growth of other members of their species
    d. the presence of chemicals in a tree's leaf and bark inhibits growth of other members of their species
    e. environmental conditions are more or less the same throughout the habitat

Use the following information for questions 7 and 8.

The sum of natality and immigration minus the sum of mortality and emigration gives a population's net growth. Natality = N, Immigration = I, Mortality = M, and Emigration = E. Each is expressed as a percent of the starting population size of 100 rabbits.

7. Which values would lead to growth of the rabbit population?
    a. N=20, I=0, M=4, E=16
    b. N=11, I=1, M=3, E=12
    c. N=2, I=15, M=12, E=0
    d. N=6, I=2, M=7, E=2
    e. N=6, I=8, M=15, E=4

8. Which values would lead to zero population growth of the rabbit population?
    a. N=20, I=0, M=4, E=16
    b. N=11, I=1, M=3, E=12
    c. N=2, I=15, M=12, E=0
    d. N=6, I=2, M=7, E=2
    e. N=6, I=8, M=15, E=4

9. How do Type I and Type II survivorship curves compare?
    a. Type I shows high mortality and Type II shows low mortality in the early years of life
    b. Both show low mortality in the early years of life.
    c. Both show high mortality in the early years of life.
    d. Type I shows increasing mortality with age and Type II shows a constant mortality.
    e. Type I shows increasing mortality with age and Type II shows decreasing mortality with age.

10. How do Type I and Type III survivorship curves compare?
    a. Type I shows high mortality and Type III shows low mortality in the early years of life
    b. Both show low mortality in the early years of life.
    c. Both show high mortality in the early years of life.
    d. Type I shows increasing mortality with age and Type III shows a constant mortality.
    e. Type I shows increasing mortality with age and Type III shows decreasing mortality with age.

Use the following information for questions 11 and 12.

R represents a population's net reproductive rate. If population size is constant, R =1. When it is greater than 1, the population is increasing in size, and when it is less than 1, the population is decreasing. Use the formula $N_1 = N_0 R$ to calculate the value of R for the growth of *E. coli* plotted in Figure 43-5 of your text. (N_1 equals population size after 1 interval of time [in this case 20 minutes] and N_0 equals the initial population size).

11. Calculate R based upon an increase from 32 (N_0) to 64 (N_1).
    a. 1
    b. 2
    c. 3
    d. 4
    e. 5

12. Calculate R based upon an increase from 2048 (N_0) to 4096 (N_1).
    a. 1
    b. 2
    c. 3
    d. 4
    e. 5

Use the following information for questions 13 and 14.

The following are age-sex structures for different populations. Assume no migration, or that immigration and emigration are equal and do not alter the age-sex structure.

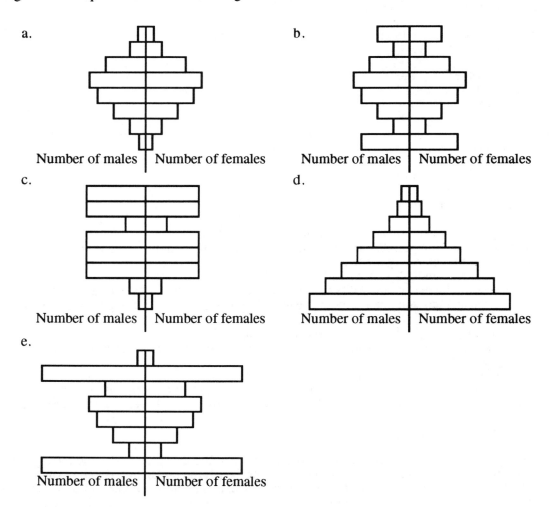

13.   Which population will grow the fastest?

14.   Which population will experience a sustained decrease in size?

15.   How are environmental resistance and carrying capacity related?
       a.   Carrying capacity determines the amount of environmental resistance.
       b.   Environmental resistance determines an environment's carrying capacity.
       c.   Environmental resistance describes the abiotic portion of an environment's carrying capacity.
       d.   Carrying capacity describes the biotic portion of environmental resistance.
       e.   They are different terms for the same thing.

16.   How do J-shaped and S-shaped population growth compare?
       a.   J-shaped growth levels off at carrying capacity and S-shaped growth does not.
       b.   Both begin with a slow rate of population increase and enter a period of exponential growth.
       c.   J-shaped growth begins slowly and S-shaped begins rapidly.
       d.   J-shaped growth enters an exponential growth phase and S-shaped does not.
       e.   J-shaped growth levels off after an initial exponential phase and S-shaped does not.

17. The Australian sheep-blowfly is used as an example of
    a. an r-selected species.
    b. a K-selected species.
    c. a species with a clumped distribution pattern.
    d. a species with a uniform distribution pattern.
    e. a species with a low intrinsic rate of increase.

18. Your text uses the extinct passenger pigeon as an example of
    a. an r-selected species.
    b. a K-selected species.
    c. a species with a clumped distribution pattern.
    d. a species with a uniform distribution pattern.
    e. a species with a low intrinsic rate of increase.

19. Which is an example of a density-dependent limiting factor?
    a. disease
    b. fire
    c. volcanic lava flow
    d. mud slide
    e. application of insecticide to a crop

20. Which is an example of a density-independent limiting factor?
    a. competition
    b. predation
    c. infanticide
    d. flood
    e. stress

21. Please examine Table 43-3. What is the historic trend of human doubling time?
    a. It has stayed pretty much the same throughout history.
    b. It increased early in human history and decreased in the twentieth century.
    c. It has generally decreased.
    d. It has generally increased.
    e. It decreased early in human history and increased in the twentieth century.

22. Please examine Table 43-3. What has happened to human doubling time in the latter part of the twentieth century?
    a. It has stayed pretty much the same.
    b. It increased, then decreased.
    c. It decreased.
    d. It increased.
    e. It decreased and then decreased again.

Concept Map Construction

Construct a concept map for each group of terms. Be sure to include appropriate connector phrases. You may add other terms as necessary and use terms in the plural or singular form.

1. limiting factor, density-dependent factor, zero population growth, natality, intrinsic rate of increase
2. carrying capacity, doubling time, tropical rain forest, biodiversity, fertility rate
3. niche, r-selected species, exponential growth, Law of the Minimum, ecological pyramid

Chapter 44

Animal Behavior

Section Concept Map

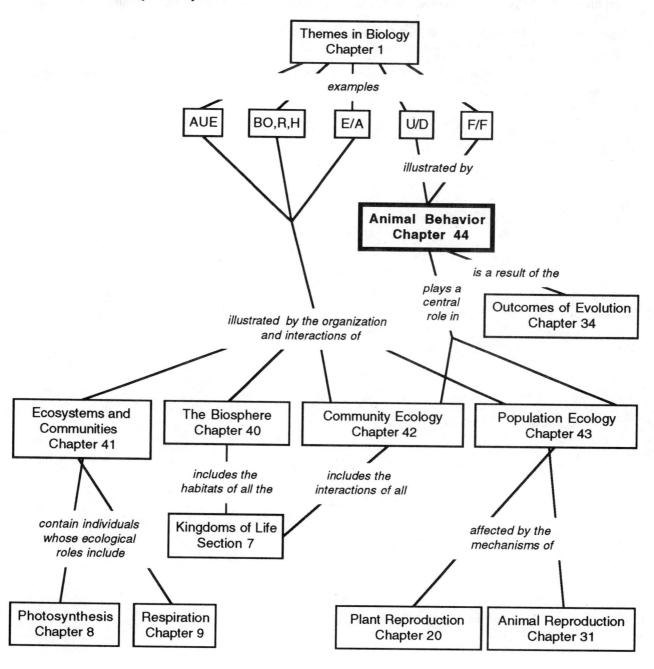

Go Figure!

Figure 44-2: Describe how you would attempt to demonstrate that the courtship of a stickleback is an innate rather than a learned behavior.

Of what selective advantage are FAPs?

Figure 44-4: Why do you suppose the saliva a dog produces was collected and measured?

Figure 44-13: How might you attempt to collect evidence to support the statement "grooming originally functioned only for skin care..."?

How might grooming behavior confer selective advantage upon the organism doing the grooming?

Matching Questions

Write the letter of the phrase that best matches the numbered term on the left. Use each only once.

_____ 1. ethology

_____ 2. innate behavior

_____ 3. sign stimulus

_____ 4. chain of reactions

_____ 5. egg-laying hormone

_____ 6. habituation

_____ 7. conditioned reflex

_____ 8. operant conditioning

_____ 9. filial imprinting

_____ 10. critical period

_____ 11. sexual imprinting

_____ 12. territoriality

_____ 13. ESS

_____ 14. mobbing behavior

_____ 15. recruitment

_____ 16. pheromone

_____ 17. altruism

_____ 18. conspecific

_____ 19. inclusive fitness

_____ 20. eusocial

a. general term for a single, species-specific, highly stereotyped, under precise genetic control animal response

b. chemical which increases the rate of action potentials generated from _Aplysia's_ abdominal ganglion

c. a learned response resulting in a new association between a stimulus and a response

d. the limited developmental period during which filial imprinting may occur

e. in young zebra finch males, a process during which they develop a courtship preference based upon the physical features of females

f. results in defense of an area against intruders

g. a group of prey intimidating and harassing a predator

h. something in the environment which illicits a fixed action pattern

i. performance of a service which benefits a conspecific at some cost to the one who performs the service

j. the study of animal behavior

k. refers to another member of the same species

l. name which applies to species that have sterile workers participate in cooperative care of young

m. the simplest form of learning

n. term which applies when individuals of a species are brought together to perform a specific task

o. requires that a behavior originally occur spontaneously

(choices continued on the next page)

p. determining an organism's adaptive value by adult offspring it leaves plus all adult offspring of relatives that survive because of the individual's actions

q. in young birds, a process of developing a preference for following their mother

r. an intricate sequence of fixed action patterns

s. a chemical signal which influences the behavior of others of the same species

t. an optimal strategy which is both "unbeatable and uncheatable"

Multiple Choice Questions

1. In general, behaviors that are precisely specified by genes are
    a. often those which must be expressed in nearly perfect form on the first trial.
    b. often those which appear in intricate patterns called a chain of reactions.
    c. associated with egg laying or mating.
    d. easily habituated.
    e. called ELHs.

2. Innate behaviors are
    a. often those which must be expressed in nearly perfect form on the first trial.
    b. often those which appear in intricate patterns called a chain of reactions.
    c. associated with egg laying or mating.
    d. easily habituated.
    e. called ELHs.

3. A fixed action pattern is
    a. a specific example of habituation.
    b. beneficial in that it eliminates a response to frequently occurring stimuli that have no bearing on an animal's welfare.
    c. beneficial in that it focuses attention and energy use on important aspects of the environment.
    d. a motor response triggered by a sign stimulus.
    e. a form of learning.

4. Classical conditioning is
    a. a specific example of habituation.
    b. beneficial in that it eliminates a response to frequently occurring stimuli that have no bearing on an animal's welfare.
    c. beneficial in that it focuses attention and energy use on important aspects of the environment.
    d. a motor response triggered by a sign stimulus.
    e. a form of learning.

5. Your text cites the retrieval of an egg by a female greylag goose as evidence that
    a. habituation occurs.
    b. behaviors are initiated by sign stimuli.
    c. once a FAP begins, it continues to completion.
    d. learning is an important part of all behaviors.
    e. unconditioned stimuli initiate operant conditioning.

6. Your text cites the retrieval of an egg by a female greylag goose as evidence that
   a. habituation does not occurs.
   b. sign stimuli are initiated by behaviors.
   c. once an FAP begins, the sequence of actions is virtually identical.
   d. learning is an important part of all innate behaviors.
   e. conditioned stimuli initiate operant conditioning.

7. When Niko Tinbergen and co-workers presented dummy male sticklebacks to real males, the researchers were able to demonstrate that
   a. sticklebacks defend their territories vigorously.
   b. a red tint on the undersurface of a model stickleback habituates real males to the presence of other real males.
   c. a red tint on the undersurface acts as a territory defense behavior releaser.
   d. the size and shape of the model acts as a territory defense behavior releaser.
   e. territory defense behavior is an example of a conditioned stimulus.

8. When Niko Tinbergen and co-workers presented dummy male sticklebacks to real males, the researchers were able to demonstrate that
   a. sticklebacks defend their territories vigorously when females are present.
   b. a red tint on the undersurface of a model stickleback conditions real males to the presence of other real males.
   c. a red tint on the undersurface does not acts as a territory defense behavior releaser.
   d. the size and shape of the model does not act as a territory defense behavior releaser.
   e. territory defense behavior is an example of an unconditioned stimulus.

9. Tinbergen and co-workers also showed that in stickleback courtship
   a. "head-up" posture acts as a releaser initiating a zig-zag courtship dance by a female stickleback.
   b. "head-up" posture acts as a releaser initiating a zig-zag courtship dance by a male stickleback.
   c. a zig-zag dance by a female initiates approach behavior by a male.
   d. a zig-zag dance by a male initiates approach behavior by another male.
   e. a female and male are never in the nest at the same time.

10. Tinbergen and co-workers also showed that in stickleback courtship
    a. "head-up" posture acts as a releaser initiating dorsal spine raising by a female.
    b. "head-up" posture acts as a releaser initiating dorsal spine raising by a male.
    c. a zig-zag dance by a male initiates dorsal spine raising behavior by another male.
    d. a zig-zag dance by a male initiates approach behavior by a female.
    e. a female and male are always in the nest at the same time.

11. Which is NOT an example of learning?
    a. habituation
    b. fixed action pattern
    c. classical conditioning
    d. operant conditioning
    e. imprinting

12. Which is NOT an example of learning?
    a. A clamworm repeatedly encounters a shadow passing overhead. At first, it quickly retreats into its burrow when the shadow appears. Eventually the worm no longer responds to the shadow.
    b. Many times food was presented to Pavlov's dogs following the ringing of a bell. Eventually the dogs would salivate when they heard the bell.
    c. A male robin entering the territory of another is vigorously attacked. A tuft of red feathers placed in a male robin's territory is also vigorously attacked.
    d. A hungry rat is placed in a box with a lever. If it presses the lever, food enters the cage. At first, the rat touches the lever while randomly exploring the box. Given sufficient time, the rat begins pressing the lever in a nonrandom manner.
    e. A gosling follows the first moving object it sees after hatching.

13. Captive chimpanzees who had no previous direct or indirect experience stacking boxes to retrieve food stacked boxes in order to reach a banana. This is an example of
    a. habituation.
    b. insight learning.
    c. classical conditioning.
    d. operant conditioning.
    e. social learning.

14. Identical twins were separated at birth. Both are now eight years old. Joey has been with their mother since he was released from the hospital and Rufus has been with their father. Neither youngster has spent any time with the other parent. Joey has a pet dog and thinks cats are for wimps; Rufus has a pet cat and is afraid of dogs. If their pet preferences were learned, it is most likely a consequence of
    a. habituation.
    b. insight learning.
    c. classical conditioning.
    d. operant conditioning.
    e. social learning.

15. One hypothesis concerning this behavior states that sensory and motor stimulation of this activity may cause the formation of cerebellar synaptic networks. What is being described?
    a. habituation
    b. play
    c. filial imprinting
    d. territorial behavior
    e. fixed action patterns

16. This type of learning is distinguished from others by the speed with which it occurs, the limited time during which it can occur, and that it occurs without any obvious reward.
    a. habituation
    b. play
    c. filial imprinting
    d. territorial behavior
    e. fixed action patterns

17. According to your text, a benefit of territoriality is
    a. reduced need to interact with others of the same species.
    b. reduced need to learn.
    c. reduced visibility to predators.
    d. reduced transmission of disease.
    e. elimination of the necessity to migrate.

18.   According to your text, a possible cost of territoriality is
       a.   increased need to interact with others of the same species.
       b.   increased need to learn.
       c.   increased visibility to predators.
       d.   increased transmission of disease.
       e.   increased necessity to migrate.

19.   According to your text, which mode of communication is best over long distances in an open
       terrestrial environment?
       a.   visual
       b.   sound
       c.   pheromones
       d.   body contact
       e.   scent trails

20.   According to your text, which mode of communication is best over long distances in the open
       ocean?
       a.   visual
       b.   sound
       c.   pheromones
       d.   body contact
       e.   scent trails

Concept Map Construction

Construct a concept map for each group of terms. Be sure to include appropriate connector phrases. You
may add other terms as necessary and use terms in the plural or singular form.

1.   innate behavior, chain of reactions, learning, habituation, unconditioned stimulus
2.   behavior, FAP, gene, transcription, ELH
3.   natural selection, altruism, kin selection, cooperative mate selection, learning

Appendix

Answers to Selected Study Guide Questions

Chapter 1

Matching Questions

1. e
3. i
5. g
7. d
9. r
11. h
13. j
15. t
17. m
19. o

Multiple Choice Questions

1. b
3. c
5. b
7. e
9. d
11. c
13. c
15. c
17. a
19. a
21. b
23. a
25. a
27. b

Chapter 2

Matching Questions

1. e
3. i
5. g
7. h
9. d

Multiple Choice Questions

1. b
3. b
5. a
7. b
8. d
9. c
10. b
11. d
12. d
19. a

Chapter 3

Matching Questions

1. n
3. s
5. j
7. i
9. d
11. p
13. k
15. l
17. c
19. f

Multiple Choice Questions

1. a
3. b
5. c
7. b
9. b
11. b
13. c
15. b
17. d
19. a

Chapter 4

Matching Questions

1. k
3. o
5. a
7. b
9. s
11. e
13. q
15. d
17. n
19. p

Multiple Choice Questions

1. a
3. c
5. a
7. b
9. d
11. c
13. a
15. d
17. a
19. e

Chapter 5

Matching Questions

1. h
3. a
5. e
7. d
9. r
11. p
13. c
15. i
17. k
19 m

Multiple Choice Questions

1. c
3. a
5. c
7. b
9. e
11. d
13. e
15. d
17. b
19. b
21. d

Chapter 6

Matching Questions

1. f
3. a
5. d
7. b
9. t
11. s
13. r
15. q
17. c
19 g

Multiple Choice Questions

1. d
3. d
5. d
7. b
9. c
11. d
13. d
15. a
17. a
19. a
21. b

Chapter 7

Matching Questions

1. o
3. n
5. b
7. c
9. f
11. g
13. h
15. m

Multiple Choice Questions

1. a
3. a
5. d
7. d
9. e
11. c
13. e
15. c

Chapter 8

Matching Questions

1. g
3. h
5. i
7. d
9. c
11. r
13. s
15. q
17. m
19. p

Multiple Choice Questions

1. c
3. b
5. a
7. d
9. d
11. d
13. e
15. d
17. a
19. e
21. a
23. b

Chapter 9

Matching Questions

1. f
3. b
5. o
7. c
9. d
11. h
13. e
15. j

Multiple Choice Questions

1. a
3. c
5. a
7. e
9. c
11. b
13. b
15. a
17. d
19. b
21. c
23. c
25. e

Chapter 10

Matching Questions

1. e
3. m
5. r
7. s
9. b
11. a
13. o
15. h
17. p
19. n

Multiple Choice Questions

1. b
3. d
5. d
7. e
9. b
11. c
13. a
15. b
17. b
19. c
21. c

Chapter 11

Matching Questions

1. f
3. a
5. b
7. d
9. n
11. j
13. g
15. k
17. t
19. s

Multiple Choice Questions

1. d
3. c
5. a
7. c
9. c
11. a
13. a
15. d
17. a
19. c

Chapter 12

Matching Questions

1. f
3. a
5. b
7. l
9. k
11. n
13. g
15. t
17. s
19. q

Multiple Choice Questions

1. a
3. d
5. a
7. a
9. b
11. a
12. c
13. c
14. d
19. c
21. c
23. a
25. c

Chapter 13

Matching Questions

1. h
3. b
5. a
7. c
9. i
10. d
11. s
15. t
17. l
19. r

Multiple Choice Questions

1. c
3. c
5. b
7. b
9. c
11. c
13. d
15. d
17. b
19. d
21. b
23. b

Chapter 14

Matching Questions

1. h
3 t
5. k
7. j
9. c
11. e
13. d
15. b
17. m
19. g

Multiple Choice Questions

1. b
3. d
5. a
7. e
9. c
11. b
13. b
15. a
17. c
18. c
19. c
20. e

Chapter 15

Matching Questions

1. d
3. n
5. o
7. l
9. c
11. m
13. g
15. f

Multiple Choice Questions

1. d
3. a
5. d
7. a
9. b
11. d
13. b
15. a
17. b

Chapter 16

Matching Questions

1. l
3. g
5. a
7. e
9. s
11. n
13. k
15. d
17. h
19. o

Multiple Choice Questions

1. d
3. e
5. a
7. b
9. c
11. b
13. c
15. b
17. c

Chapter 17

Matching Questions

1. j
3. h
5. l
7. i
9. o
11. f
13. c
15. g

Multiple Choice Questions

1. d
2. e
3. c
7. a
8. c
9. c
13. b
14. c
15. a
16. e
21. d
23. c

Chapter 18

Matching Questions

1. h
3. j
5. r
7. m
9. b
11. i
13. g
15. l
17. c
19. t

Multiple Choice Questions

1. a
3. d
5. b
7. d
9. a
11. c
13. e
15. d
17. c
19. b
21. d
23. a
25. d

Chapter 19

Matching Questions

1. g
3. l
5. m
7. n
9. c
11. b
13. f
15. h

Multiple Choice Questions

1. d
3. c
5. c
7. b
9. b
11. d
13. d
15. b
17. c
19. c

Chapter 20

Matching Questions

1. g
3. k
5. d
7. t
9. r
11. e
13. b
15. q
17. i
19. c

Multiple Choice Questions

1. c
3. b
5. c
7. d
9. d
11. d
13. d
15. b
17. c
19. a
21. c

Chapter 21

Matching Questions

1. f
3. p
5. m
7. q
9. j
11. g
13. h
15. a
17. c

Multiple Choice Questions

1. b
3. a
5. b
7. e
9. a
11. c
13. c
14. d
17. e
18. d

Chapter 22

Matching Questions

1. h
3. j
5. g
7. b
9. e
11. f
13. o
15. p
17. m

Multiple Choice Questions

1. b
3. c
5. a
7. a
9. c
11. c
13. c
15. b
17. c
19. a

Chapter 23

Matching Questions

1. j
3. a
5. h
7. o
9. c
11. f
13. q
15. m
17. e
19. n

Multiple Choice Questions

1. a
3. d
5. c
7. d
9. b
11. e
13. b
15. b
17. c
19. e
21. c

Chapter 24

Matching Questions

1. l
3. r
5. b
7. t
9. e
11. d
13. p
15. g
17. m
19. o

Multiple Choice Questions

1. c
3. d
5. d
7. c
9. a
11. b
13. e
15. c
17. a

Chapter 25

Matching Questions

1. s
3. t
5. m
7. a
9. p
11. h
13. o
15. n
17. r
19. g

Multiple Choice Questions

1. c
3. b
5. d
7. a
9. b
11. a
13. d
15. c
17. b
19. b
21. e
23. b

Chapter 26

Matching Questions

1. g
3. i
5. l
7. n
9. c
11. e
13. d
15. q
17. p
19. f

Multiple Choice Questions

1. c
3. a
5. b
7. b
9. b
11. a
13. b
15. d
17. d
19. c
21. a
23. c

Chapter 27

Matching Questions

1. c
3. f
5. a
7. l
9. s
11. h
13. e
15. d
17. r
19. k

Multiple Choice Questions

1. d
3. d
5. a
7. d
9. b
11. b
13. c
15. b
17. c
19. b

Chapter 28

Matching Questions

1. i
3. g
5. a
7. b
9. q
11. t
13. c
15. e
17. h
19. j

Multiple Choice Questions

1. d
3. b
5. b
7. b
9. b
11. b
13. a
15. b
17. b
19. d
21. c
23. a
25. b

Chapter 29

Matching Questions

1. k
3. p
5. b
7. a
9. t
11. c
13. r
15. q
17. l
19. o

Multiple Choice Questions

1. a
3. a
5. c
7. d
9. a
11. a
13. c
15. e
17. b
19. d
21. c
23. a

Chapter 30

Matching Questions

1. i
3. k
5. b
7. f
9. c
11. l
13. s
15. g
17. e
19. h

Multiple Choice Questions

1. a
3. e
5. d
7. d
9. b
11. e
13. c
15. b
17. d
19. c

Chapter 31

Matching Questions

1. d
3. i
5. a
7. f
9. t
11. c
13. k
15. p
17. n
19. h

Multiple Choice Questions

1. c
3. b
5. c
7. c
9. a
11. a
13. a
15. c
17. d
19. c
21. b
23. a

Chapter 32

Matching Questions

1. i
3. p
5. t
7. m
9. s
11. f
13. l
15. q
17. j
19. k

Multiple Choice Questions

1. c
3. c
5. d
7. b
9. c
11. e
13. b
15. c
17. d
19. a

Chapter 33

Matching Questions

1. i
3. b
5. d
7. n
9. q
11. h
13. m
15. e
17. k
19. c

Multiple Choice Questions

1. a
3. d
4. a
7. a
9. a
11. d
13. a
15. b
17. a
19. b

Chapter 34

Matching Questions

1. f
3. a
5. l
7. g
9. b
11. d
13. p
15. o

Multiple Choice Questions

1. d
3. b
5. a
7. c
9. c
11. b
13. c
15. e

Chapter 35

Matching Questions

1. i
3. j
5. c
7. o
9. f
11. h
13. e
15. g

Multiple Choice Questions

1. d
3. d
5. a
7. e
9. a
11. e
13. b
15. d
17. a
19. b

Chapter 36

Matching Questions

1. h
3. j
5. m
7. g
9. b
11. p
13. r
15. i
17. f

Multiple Choice Questions

1. e
3. b
5. d
7. d
9. c
11. b
13. a
15. a
17. d
19. b

Chapter 37

Matching Questions

1. e
3. m
5. b
7. o
9. n
11. g
13. l
15. i

Multiple Choice Questions

1. d
3. d
5. e
7. a
9. b
11. e
13. e
15. b
17. c
19. b

Chapter 38

Matching Questions

1. j
3. a
5. q
7. d
9. c
11. e
13. t
15. f
17. n
19. h

Multiple Choice Questions

1. c
3. b
5. e
7. d
9. c
11. c
13. a
15. b
17. a
19. b
21. c
23. b

Chapter 39

Matching Questions

1. d
3. t
5. o
7. r
9. m
11. a
13. f
15. g
17. l
19. i

Multiple Choice Questions

1. a
3. c
5. e
7. c
9. d
11. e
13. a
15. b
17. a
19. c
21. c
23. d
25. b
27. a
29. a

Chapter 40

Matching Questions

1. j
3. f
5. s
7. b
9. m
11. r
13. q
15. n
17. c
19. i

Multiple Choice Questions

1. a
3. c
5. c
7. c
9. b
11. b
13. c
15. c
17. a
19. d
21. d

Chapter 41

Matching Questions

1. k
3. t
5. o
7. a
9. m
11. q
13. d
15. r
17. p
19. g

Multiple Choice Questions

1. a
3. c
5. d
7. a
9. a
11. b
13. b
15. a
17. d
19. c
21. c

Chapter 42

Matching Questions

1. i
3. b
5. s
7. t
9. d
11. o
13. r
15. g
17. n
19. m

Multiple Choice Questions

1. c
3. b
5. b
7. d
9. e
11. b
13. a
15. a
17. a
19. d

Chapter 43

Matching Questions

1. d
3. r
5. e
7. b
9. k
11. l
13. j
15. f
17. m
19. o

Multiple Choice Questions

1. c
3. c
5. e
7. c
9. d
11. b
13. d
15. b
17. a
19. a
21. c

Chapter 44

Matching Questions

1. j
3. h
5. b
7. c
9. q
11. e
13. t
15. n
17. i
19. p

Multiple Choice Questions

1. a
3. d
5. c
7. c
9. b
11. b
13. b
15. b
17. d
19. c

NOTES

NOTES

NOTES

NOTES

NOTES

NOTES

NOTES

NOTES

NOTES

NOTES

NOTES

NOTES